War Narratives

Number 163
Williams-Ford Texas A&M University
Military History Series

War Narratives

SHAPING BELIEFS, BLURRING TRUTHS IN THE MIDDLE EAST

Caleb S. Cage

TEXAS A&M UNIVERSITY PRESS
COLLEGE STATION

First edition

This paper meets the requirements of ANSI/NISO Z39.48–1992 (Permanence of Paper).
Binding materials have been chosen for durability.
Manufactured in the United States of America

Library of Congress Cataloging-in-Publication Data

Names: Cage, Caleb S., 1979– author.
Title: War narratives: shaping beliefs, blurring truths in the Middle East / Caleb S. Cage.
Description: First edition. | College Station: Texas A&M University Press, [2019] | Series: Williams-Ford Texas A&M university military history series; number 163 | Includes bibliographical references and index. | Identifiers: LCCN 2018047513 (print) | LCCN 2018050927 (ebook) | ISBN 9781623497613 (ebook) | ISBN | ISBN 9781623497606 (cloth: alk. paper)
Subjects: LCSH: Iraq War, 2003–2011—Propaganda. | Afghan War, 2001—Propaganda. | Iraq War, 2003–2011—Public opinion. | Afghan War, 2001—Public opinion. | Information warfare—United States. | United States—Military policy. | War on Terrorism, 2001–2009, in mass media. | War in literature.
Classification: LCC UB276 (ebook) | LCC UB276.C34 2019 (print) | DDC 956.7044/31—dc23
LC record available at https://lccn.loc.gov/2018047513

For my father, Gary Cage,
who has always taught me to pursue the truth,
and who has shown me how to do so
through the way he lives.

Contents

Foreword

THE KAYAK tumbles down a rapidly moving tributary, while its navigator surveys the shore for a spot suitable for another night of camping. Unbeknownst to him, a small biplane zooms through the canyon. The pilot had just come over the crest of the mountain range and had witnessed small streams emerging from the creases in the mountainside, then forming the kayaker's tributary. The pilot then flies to where the tributary joins a river that eventually empties into a large lake.

In the words of philosopher Louis O. Mink, "Stories are not lived; they are told." In terms of the kayaker, events in the world occur in a specific period of time, but the narrative that tells their story is told from the perspective of the pilot, who can see the events leading up to the central ones as well as see the results to which those same central events led. That is, a narrative is always told from the perspective of the pilot, while the events of the narrative are always lived from the perspective of the kayaker.

This transcendent quality is the essential aspect of a narrative, giving meaningful context to a series of brute, raw facts. Narratives, therefore, are historical tapestries, and as such, they are works of art. Narratives are creations from transcendence, from the perspective of the pilot flying along the ridge of the mountains. Like the story of Genesis in the Bible, the events are void and without form until the spirit of the storyteller hovers over them to give them order and shape. The creator of the narrative selects those events, and aspects of events, that give the greatest force to the basic premise of the story. There is a theme and a purpose in his mind before he starts the story, and that theme is to his narrative as a theory is to a scientific experiment.

As noted, narratives are artful tapestries, but perhaps an even better metaphor is music, as Lydia Goehr asserts. Goehr compares a narrative to a musical composition of several movements. That is, a good narrative places events in a certain order that creates anticipation, excitement, and even resolution, all reflecting the transcendent perspective of the composer.

This transcendent quality is also significant in another aspect. Narratives can set out the development of unintended consequences. Ironically (as we shall see), Hegel was one of the first to emphasize these unintended and unexpected consequences in history. These consequences are a part of the storyline, and as such, can be included into the arc of the narrative as the story continues.

History is replete with such cases. The Roman church sought to devastate the Byzantine Empire by its sacking of Constantinople in 1204, but as a result, the scholars and artists of that city fled to northern Europe from where they laid the groundwork for the Renaissance, which later toppled the Holy Roman Empire. The Black Plague of fourteenth-century Europe served to launch the greatest expansion of medical science in all of history. General Lee, on the morning of July 3, 1863, intended his attack on Seminary Ridge to bring the war to a rapid conclusion. Instead, the attack turned the war against Lee and inaugurated a grinding decline for the South.

Ivan Illich argues, in fact, that it is the nature of human institutions, not only to fail in solving the problems for which the institutions were created, but instead to exacerbate that problem. Thus, many institutions, started with the best of intentions, are found later, according to Illich, to have produced unintended and undesired consequences. Because of this aspect, one may ask whether Luther indeed opened the door to religious liberty, maybe even liberty in general, even though he intended only to correct certain abuses in the church. Or did Nietzsche's philosophy lead to Nazism? These questions are often debated. Narratives proffer answers.

Yet another quality of good narratives, which bespeaks of transcendence, is the narrative writer's awareness of the strangeness of the past—not only of his/her past, but also of other cultures. In the times before airplanes, the highest crests of the mountains were also the highest vistas. Only rarely was a person from these times able to see his region from a heightened perspective.

Likewise, in a shame/honor society a person might, out of a sense of honor, perform deeds that people in other cultures would regard as crimes. Narratives that include such times and cultural aspects will be told in the tropes of the narrator, and of his hearers as well. And the narrative writer always relates the events in his own terms.

For example, one might explain the fall of the Roman Empire in terms of modern economic principles. Or the modern narrativist might choose to describe Nero's behavior in Freudian psychological terms. The narrativist will also relate the events of his story to the concerns of his own day and audience. But his respect for the differences between different times and different cultures reflects his transcendent position; otherwise, he would not see those differences.

That is what we mean by transcendence: that artistic description of a series of events from a perspective higher than the players. But to say that narrative is like art is not to say that historical understanding is relativistic. In fact, it is quite the contrary. Just as a piece of art might more or less represent reality, a narrative might more or less represent not only what happened but also what it all meant—and in fact, it might get it right. Moreover, this implies that those who later either related or heard a certain narrative might very well have understood a certain event better than even those figures who actually lived the event.

That also means, of course, that there are poor narratives. There are revisionist accounts, accounts with agendas, political, personal, or otherwise, and there are times when historians make mistakes. It is hard to tell what the historical figures actually believed or what their motives were. This, in fact, is the hardest part of a narrative. And even if the players have testified to their beliefs and motives, there is always the possibility they were lying—or worse, that they were unaware of their true beliefs and motives.

As philosopher Thomas Nickles has shown, even in the sciences people often believe they were thinking one thing and interpret their own work accordingly, even though it can be shown that that was not the case. This is so common, in fact, that it has a name, "the Whig interpretation of history." Freud, Nietzsche, and Marx are famous for their suspicions that the reigning narratives had been originated with ulterior motives. One might also fabricate a story to justify actions, whether past, present, or future, and might even, after the repeated telling of it, come to believe it himself.

Narratives, therefore, like art, can be used for political propaganda. We see this use in current Islamic fundamentalist terrorist groups, though this use is not exclusive to them, certainly. Their followers are told a horrific story of abuse by westerners, a story, though true in some respects, that is told in a biased fashion in order to create a profound sense of desperate victimization.

Fortunately, though, narrative accounts can be adjudicated, and some argue that this begins with resisting grand narratives. Eusebius's *Church History* is a good example of a master narrative, one in which Constantine is seen as raised up by God to free the Christian church from persecution and put it on a course for European rule. Hume's *History of England* is also important because he rejected the concept of a master narrative. We should be wary of those narratives that insist that history is moving inexorably toward a certain goal, that there is a manifest destiny at some point ahead, or there are those signs that can be read in the events of history. Such would be an extraordinary claim and would require extraordinary evidence.

But on the other hand, neither do we need to resort to postmodernism. Not every account is as good as the other. Some historical accounts are wrong. We can adjudicate narratives, compare them, and assess them for their approximation to the truth. Checking for consistency, taking the storyteller's agenda into account, and finding physical evidence can help ascertain correctness in a narrative. Multiple narratives from different perspectives, even narratives with competing agendas, are invaluable with regard to finding the truth about what happened in the past. But the result will always be incomplete, imperfect, and itself another narrative subject to critique.

But we return to the transcendent aspect of narratives. On the morning of May 23, 1618, certain Protestant barons of Bohemia defenestrated their Catholic overlords from the third floor of Prague Castle. The overlords survived the fall, but the event constituted a clear statement of Czech Protestant defiance against Catholic control. However, no one involved in that upper story brawl would have been able to say that day that said event started the Thirty Years' War. How could anyone there that day possibly know that Catholic bishops would send a German military force (including the young René Descartes) to quell an uprising of Bohemians and Moravians, now known as the Battle of White Mountain. The mothers of those three hundred brave Moravian warriors, who stood there and died to a man rather than allow their opponents to think for even a moment that they were not determined, could not have known that eventually they would be able to say that their sons had not died in vain. And no one would have known that the Protestant king of Sweden, Gustavus Adolphus, would enter the fray to turn the tide of the war, leading to the Peace of Westphalia in 1648, which at once would establish a place for Protestants in Europe and introduce a long period of peace in Europe. But the modern historian knows these things. She sees the tributary in both directions. She sees now how the Thirty Years' War started and what it led to. She also can derive meaning from that perspective.

In fact, there is no meaning without narrative. A husband who loses his wife to Alzheimer's disease has lost a shared narrative. What is a relationship other than a history, which is known by the partners, shared constantly and used to explain who they are and why they do what they do? It was once thought that all human knowledge, including history, could be subsumed under the rubric of empirical science. In fact, a dream of modernism was that all of human understanding might be brought into a unity, and it was presumed that the bonding element would be empirical science. It was expected that psychology, history, sociology, and even theology would be brought under the aegis of science. But philosophers Norman Hanson and Thomas Kuhn changed all of that. We now think that a better rubric is history. Recently, philosophers such as William Walsh, William Dray, Morton White, Arthur C. Danto, and the leader of them all, Hayden White, have come to see narrative as the essential nature of history. Understanding anything is always in terms of a narrative. Not only that, but every field of inquiry should be subsumed under history, even the sciences. When a young student is being trained in the field of biochemistry, she is actually being drawn into the narrative of that discipline. She enters the chain of scientific discoveries at its present point, but all of her training is built on the narrative of that discipline, which is constantly told and retold and used to explain her present researches. A historical narrative, then, is everyman's philosophy. It combines empirical facts, historical chronology, and philosophical interpretation—perhaps even theology, as well.

Philosopher Arthur Danto warns of a dangerous conceit with regard to narratives. There are three asymmetric modes, he says, namely, the past, the present, and the future. The dangerous conceit lies in thinking that historical words uttered after an event can also, by some sort of extension, be uttered before them. It would be a conceit for someone to see himself as a George Washington redivivus, and a dangerous conceit at that. This is evidenced in the many dictators who think they know in advance what the consequences of their actions will be and that they know what is best for humanity. We also live in a constantly moving present, and there is the temptation to interpret past events in terms more relevant to the present time than those past event in fact were. On the other hand, the present of the narrative writer is indeed relevant in that the significance of the original events can change with time. Again, this is not a crude relativism but rather a healthy recognition of perspective. Someone who was enjoying peace and freedom in 1700 might thus describe the Thirty Years' War with a particular relish.

There must be, therefore, a balance between facts and interpretation. As Lydia Goehr says, facts without interpretation are plain, but interpretation without facts is fiction. Good historical narratives, then, are those that describe events the way they were and apply their significance to a certain group of hearers. This can never be done perfectly and completely, because no narrative writer will ever know all the variables. The pilot in the biplane sees a great deal but far from everything. Therefore, not only should the storyteller maintain the appropriate humility, which comes from her lack of omniscience, but the hearers need to recognize the essential open-endedness of all stories. No one can ever know the ultimate significance of any series of events. Again, this is why we should be wary of grand narratives. Similarly, as philosopher Karl Popper has shown us, no ordinary narrative writer should want to assume that his story reflects knowledge of some grand Hegelian-like theme (as in Marxism or apocalypticism).

But that said, good narratives do spare us from historical naïveté. Plants and animals endure events and the passage of time, but it is distinctively human to tell stories, just as it is distinctively human to find meaning. A good historical narrator should want to get the facts right. He respects the differences between peoples, and even individuals, and struggles to relate those differences to his hearers. A good narrator might at times surprise his hearers with unexpected outcomes. And he remains humble in the awareness that his story cannot tell it all and very likely will one day be out of date. But in the telling, the story helps us understand the greatest truths about ourselves.

—Gary T. Cage, PhD

Acknowledgments

I AM GRATEFUL for the help from many over the years that it took for me to write this book. They were readers, reviewers, supporters, friends, and family, and some likely did not even know that they were helping me complete this project. Their support came in countless forms, and I simply could not have completed this work without them.

To begin, I would like to express my deepest gratitude to those who have served in the wars in Iraq and Afghanistan since 2002, and to their families. In the pages that follow, I am critical of aspects of these wars; however, I remain inspired by the commitment to service and sacrifice that I have seen from so many who willingly deployed and tried to make things better. I am proud of my own service in Iraq, and proud to have served with so many of my heroes, and this book is written with the highest respect for them.

During the latter phases of writing this book, I completed a graduate program at the Naval Postgraduate School's Center for Homeland Defense and Security. Not only was I regularly challenged on my own deeply held beliefs about national security, but my ability to see related issues from varying and complex perspectives was also greatly improved. I am especially grateful to Professors David Brannan, Christopher Bellavita, and Carolyn Halladay for their commitment to the studies of national and homeland defense, and for all they did to grow and challenge me personally.

Many of my dear friends encouraged me to finish this project, no matter how impossible it seemed at times. I am grateful for the support of Dale Erquiaga, Gregory Tomlin, Cyndy Ortiz Gustafson, Travis Bow, and many others who read early versions of this book, who reviewed and offered feedback, or who encouraged and otherwise supported me during this process. I am particularly thankful for Rob McDonald, my dearest friend, who has patiently discussed these ideas with me for many years.

Thanks goes as well to the excellent team at Texas A&M University Press, who shepherded this project through the review and publication processes. Many worked behind the scenes throughout the development of this manuscript, but I am particularly appreciative for Dr. Jay Dew, the editor in chief at the press. It is no exaggeration to state that this book would not have happened without his leadership and support.

Many of the chapters in this book were published in some form elsewhere leading up to publication. Versions or portions of chapters presented here were published in *Small Wars Journal*; *War, Literature, and the Arts*; *Michigan War Studies Review*; *Veterans Law Review*; and elsewhere. The editors of these publications not only graciously published these various works, but they also improved my writing through their feedback and edits.

Without question, my greatest thanks go to my family. My parents, Gary and Charlotte Cage, taught me to ask hard questions, to challenge my beliefs constantly, and to pursue the truth but to always be open to revising my beliefs. There is no question that they are the soul of this work, and I am grateful to them for eternity. My brothers Zach and Charlie, and my sister, Khalilah, have always been strong supporters of my work and have provided enormous encouragement through the years. It should go without saying that my dear wife, Brooke N. Cage, and my children, Charlotte N. and Arlo G. Cage, have provided the love, support, and laughter to last a lifetime.

War Narratives

Introduction

War Narratives

WHAT ARE YOUR deeply held beliefs and conclusions about the recent wars in Iraq and Afghanistan? How does reflecting on these wars make you feel? How did you get your information in developing these beliefs or drawing these conclusions? And do you hold opinions on these wars that do not align well with your political or cultural worldviews?

People hold well-informed beliefs about these wars that differ from others, who also hold well-informed beliefs. Some feel anger, discouragement, fear, and other negative emotions when they reflect on the causes and costs of these wars, while others feel pride, accomplishment, and perhaps even the fulfillment that comes from sacrifice and service. Of course, an entire spectrum of thoughts and feelings about these wars exists beyond these examples, and they are held, endured, or fostered either by people who participated directly in the war, were close to those who did, or even by those who were completely disconnected from the war effort.

The answers to these questions are your own, and this book does not attempt to invalidate any beliefs or feelings represented on this broad spectrum of personal responses to the wars. This book is not a scholarly work of sociology, political science, or even rhetoric, all of which are appropriate assumptions given that the primary subject matter is the narratives surrounding the wars in Iraq and Afghanistan. It does not delve deeply into the literature of media studies to determine how certain beliefs became conventional wisdom while others did not. And it does not pretend to provide definitive conclusions on which beliefs and feelings one ought to embrace, and which should be rejected.

Rather, this book begins with the basic assumption that much of what is considered to be conventional wisdom about these wars is based on narratives. These narratives inform various perspectives, which accounts for the diversity of opinions about the invasion of Iraq, the level of attention and support the war in Afghanistan received, or who should be held accountable for various decisions and failures. And the narrative a person embraces in developing their conclusions about these wars often align with their political, social, and cultural worldviews.

Based on these assumptions, this book attempts to show that narratives play a significant role in shaping how these wars were viewed by various constituencies at any given point during the wars, and that this even continues today. If narratives played a powerful role in meaning making for these wars, a critical observer could be forgiven for asking who was benefiting from certain messaging that contributed to the development of a given narrative, especially in cases where political leaders promoted one narrative over another. If not convinced by what they found, this same observer would be completely justified in digging deeper for the truth about these wars, seeking a truth that goes beyond comfortable confirmations of what they already believe about politicians, generals, the enemy, or a whole list of others who made decisions that resulted in the outcomes of these lengthy and costly wars.

The wars in Iraq and Afghanistan are no different because they offer fresh examples of how wars are interpreted through complex lenses by the societies that fight them, but the factors that contributed to these interpretations are unique to these wars. For the first time in the nation's history, the nation's wars were fought by an all-volunteer force, a fact that has created an enormous divide between the civilian public and its military. Further, these wars have taken place during the information age, allowing the heavily mediated information sources of new media and cable news to change the way the wars are conveyed to the public, which has influenced how they are understood by the same. These factors and others have created many flawed narratives that merely explain the wars in terms of accepted political and cultural worldviews and perspectives.

Narratives are not just the stories shared about the wars, or even how they are told, as a common usage of the term implies. Greater than just the structure, character, and themes of a story, narratives reflect the choices that the narrator makes in the ordering of his or her facts, purposely or not. They also reflect the choices the narrator makes in including or excluding certain facts in their presentation. That is to say, the way a story is told relates to how a story is understood, whether the narrator intends to influence the reader, listener, or observer or not.

Because of the new factors just listed, narratives surrounding the wars in Iraq and Afghanistan have led many to simply embrace conclusions that align with their own political and cultural worldviews, instead of investigating the causes and realities of these wars more deeply. This has resulted in a highly emotional, divided, and partisan understanding of these wars, and worse, if these divisive narratives are not challenged and critiqued, then they could potentially continue to divide the nation and inform future national security decisions.

The various narratives of these wars are continually evolving, just as the security situations in Iraq and Afghanistan evolve to this day. However, snapshots of these narratives are collected in the books representing the stories of the wars. Memoirs, journalistic accounts, oral histories, novels, and others works capture different versions of significant events, sometimes, perhaps, attempting to preserve a specific story for posterity, attempting to settle scores, attempting to build legacies, or even some combination of all of these. Whatever their stories or their motivations, though, these accounts have blurred the collective truths about the wars, or at least prevented an independent and honest pursuit of them.

Many detractors of the George W. Bush presidency, and especially his wars in Iraq and Afghanistan, believe in a three-part narrative about his administration. First, they argue that he alone took the nation to an optional war in Iraq after abandoning his early successes in Afghanistan; second, they argue that his unilateralist approach and cowboy persona continued into his approach to foreign policy, leaving the nation isolated and almost ensuring defeat; and third, that he was surprised by the turmoil his military encountered in Iraq after the invasion, which was proven by his apparent lack of planning for the occupation and reconstruction efforts in the country. While President Bush himself even embraced aspects of these narratives, there is tremendous evidence suggesting that the truth is far more complex than simply blaming a single president or his administration for the disastrous outcomes in Iraq, and even Afghanistan.

President Bush and his administration not only embraced certain aspects of these narratives at times, as seen through his tough foreign policy rhetoric, they also enjoyed the benefits of narratives later in the war. Primarily, the Bush administration benefited from the widely held belief that, after years of failure in Iraq, General David Petraeus and his surge strategy saved the nation from defeat at the hands of the insurgency in Iraq. Again, the facts do not support this narrative, as even many proponents of David Petraeus and the surge narrative admit in their works.

There are other narratives of these wars as well, all aligning with political beliefs and worldviews of a divided nation that is too often too distant from the war effort to know it firsthand. The best challenges to them come from words of those who experienced wartime deployments and combat themselves. These accounts challenge the narratives directly, they provide a helpful master narrative for understanding them better, and they often aim to deepen the conversation about the wars with the broader civilian public.

War Narratives examines these narratives individually through distinct collections of literary works that cover an equally distinct period or idea from the wars: memoirs of members of the Bush administration, memoirs based on blogs, works based on oral histories, and the literary fiction coming from these wars, among others. By doing so, *War Narratives* attempts to show how the narratives were used by those who stood to gain from them, mostly politically, and points out their inherent contradictions. It attempts to achieve a delicate balance by pointing out flaws in common wartime narratives without merely offering counternarratives that align with an opposite political or cultural worldview.

In examining these narratives individually, *War Narratives* aims to explore the stories of these wars in a way that will result in a more authentic and less manipulated way. Just as it does not attempt to show how one narrative is flawed by offering an equally flawed counternarratives, it does not reject facts and significant storylines merely because they align with a narrative that achieved or prevented a certain political outcome. Rather, *War Narratives* accepts that the information space is heavily mediated by narratives, yet one can achieve a more nuanced understanding of these wars by examining how these narratives interact with one another. After all, just like the media landscape they are interpreted through, wars are extremely complicated and dynamic, and the truth is seldom clear.

In examining the interplay of the narratives that describe these murky events, *War Narratives* argues that the conventional wisdom about these wars is often based on oversimplifications. While being broadly for the wars, against the wars, or proud or fearful might be personally comfortable, at least inasmuch as these beliefs or feelings align with one's already deeply held beliefs, these basic conclusions are not helpful in understanding these wars or understanding the nuances of future wars. Real attempts must be made, then, to have a real understanding of what brought the wars about, what led to the many failures that occurred throughout their execution, and the real impact these wars have had and will continue to have on the nation.

War Narratives pursues this argument largely by considering individual narratives and attempting to show the complexities behind their oversimplifications. It begins broadly, pitting an early narrative developed by President Bush after the September 11 attacks against its persistent counternarrative prominent in the communications of radical Islamic extremists, and then it moves through a series of four associated narratives about decision-making and policy making at the highest level of American government. Then it explores narratives on the other end of the spectrum, the narratives told about and by American service members about their experiences.

Given this introduction of the topic, it may seem easy to conclude that *War Narratives* favors the perspectives of the individual service member presented in the latter half of the book over those of the policy makers examined in the former. This is true, but only to an extent. Taken in its entirety, *War Narratives* attempts to make the case that understanding the truth about all levels of these wars is critical, as is supported by the final chapter, which argues that the most trustworthy perspectives from these wars have come from fiction written by both veterans of these wars and civilians.

The opening chapter, "Democratization and Nation Building," is built around a study of "the Narrative" in extremist Islam, and how that narrative works against President Bush's attempt to establish the narrative that the nation was attacked because extremists "hate our freedoms." The "Narrative" of extreme Islam essentially states that, by invading Iraq and Afghanistan, the West and its allies were merely continuing in a historic effort to destroy Islam, capture and control the valuable natural resources of the Middle East, and humiliate all Muslim people. It is a powerful and effective tool for extremists, allowing its purveyors to both recruit and motivate new members while also justifying their tactics, which often result in death and destruction for their fellow Muslims.

"Democratization and Nation Building" analyzes the way the Narrative works in extremist rhetoric and literature and asserts that others have misapplied the power of the narrative by assuming groups like Al-Qaeda and the Islamic State, or ISIS, have the same strategic goals. It argues that these groups are in fact tremendously different in several ways, but most important is that ISIS appears to be pursuing the establishment of a premodern caliphate, while Al-Qaeda is pursuing a postmodern movement of perpetual global jihad. Both militant groups invoke "the Narrative" to achieve these goals by eroding the modern political and religious barriers, respectively, they confront. This chapter also analyzes the scholarship around the Narrative in order to provide a study of how wartime narratives are seen to work, which is intended to be useful in the subsequent examinations of Western narratives in the chapters that follow.

The next four chapters of *War Narratives* explore the complexities of four distinct but connected narratives that figure prominently in the American understanding of the wars in Iraq and Afghanistan. These are primarily narratives about the decisions on policies and actions made at the highest levels of the Bush administration during the war years. And these chapters explore key works published from and on the period to expose challenges that may undermine the narratives.

"The Deciders" examines the war of choice narrative, which has been central to a fundamental criticism of the Bush administration by its opponents. It asserts that President George W. Bush was alone in choosing to take the nation to war in Iraq, and he is therefore alone in bearing the blame for its disastrous results. By referencing the memoirs of key Bush administration officials from this period, "The Deciders" examines three factors that complicate the war of choice narrative: presidential war powers, congressional oversight of military actions, and the role of the all-volunteer military. In doing so, this chapter shows how Bush administration officials rejected the negative aspects of the unilateralist charge central to the war of choice narrative while also embracing the narrative's more favorable aspects. It also describes a process through which many were involved in the decision to go to war, but in such a passive way that the only people who truly decided to go to war were the members of the all-volunteer military force.

"Cowboy George's Painless War" examines the cowboy narrative, which criticizes the Bush administration for not only being willing to "go it alone" into Iraq, as in the case of the war of choice narrative, but also for the administration's willingness to commit the nation to total war all over the globe. By relying on the four books written by Bob Woodward about the Bush administration at war to provide a framework, this chapter shows that while Bush was eager to exhibit tough rhetoric, he was not interested in mobilizing the nation in any serious way. This divide between talk and action is particularly apparent in Bush's constant drive to have "boots on the ground" quickly in Afghanistan and Iraq, but also in his complete unwillingness to take casualties or call for other national sacrifices at least until the surge in 2007. The same divide provided him space to push his policies further, but it also ensured an impossible tension that limited his will and his ability to fully carry out his goals.

The next chapter, "A Phase IV Rodeo," addresses the Phase IV narrative, which asserts that the Bush administration never had a plan for the reconstruction phase after the invasion in Phase III of the military plan. However, an examination of the limited number of works by those directly involved in the reconstruction effort in Iraq, and particularly *Hard Lessons*, the report of the Special Inspector General of Iraqi Reconstruction, it is clear that there was ample Phase IV planning underway before the invasions. What was missing, however, was a coherent vision for success, leadership, and coordination that would ensure a unified planning process. These missing elements set the foundation for what followed during the long and arduous reconstruction effort in Iraq: an ongoing disaster hampered by continuous violence, political squabbling between powerful government agencies, massive and reactive course corrections to the coalition's strategy, and in the end, a failed Phase IV effort.

"COIN Toss" challenges the surge narrative, a prevailing narrative of the Iraq War, and to a lesser extent, to the Afghanistan War. This narrative asserts that through late 2006, the war in Iraq was failing due to poor leadership from the effort's top general officers, but by February 2007, the war effort turned around due to the leadership of General David Petraeus and the introduction of more combat troops through the surge. The surge narrative can be seen as the Bush administration's response to the culmination of the previous three narratives into a successful, powerful, and pervasive master narrative, whether it was intentionally deployed for that purpose or not.

The surge narrative asserts that the surge was a major military success, and it is based on the proposition that everything that happened in Iraq before Petraeus was a major failure. By examining works that promote or are based on the surge narrative, this central element of the narrative is called into question. When historical military accounts of the surge, and of counterinsurgency, or COIN, are also considered, it appears that the surge might be better seen as a domestic political success than a strategic military success.

In the chapters that follow "COIN Toss," the wars in Iraq and Afghanistan are examined through the various literature produced directly after the wars. These chapters look at the war from the perspective of the individual service member as presented through oral histories, memoirs that are stylistically unique to these wars, and the growing body of wartime literary fiction from veterans and civilians of the Iraq and Afghanistan war era. These chapters examine persistent narratives assigned to veterans of these wars, provide a coherent master narrative for the other wartime narratives to reference, and try to find common meaning and purpose for the literary fiction coming from veterans and civilians through interviews, respectively.

"Heroes and Victims" provides analysis of two competing narratives that often drive the broader understanding of veterans returning from these wars, the hero narrative and the victim narrative. The hero narrative, which is more often preferred by the politically conservative, proclaims categorically that all who served in the military are heroes. The victim narrative, which is broadly a liberal position, sees those who served in the military during these wars as broken, victims of the president's folly who are in desperate need of support and services from the civilian community that sent them. While favored by their respective political ideologies, both narratives are common across the political spectrum, and neither is helpful in understanding actual wartime experiences.

Oral histories provide a way to test these narratives against how veterans actually describe their own experiences. By examining works based on oral history accounts, this chapter finds that some authors remain committed to preserving elements of the hero and victim narratives, while one work in particular provides a considerably more complex perspective, one that shows veterans rejecting these narratives directly. In it, readers can also see veterans coming to terms with their experiences personally, sometimes during the interview itself, and finding real and personal growth from their combat deployments and experiences.

"On Chickenshit" examines a military phrase used more commonly in previous wars, but that continues to describe timeless aspects of military life. Chickenshit, according to historian Paul Fussell about World War II, describes the members of the military using their legal authority to inflict petty and superficial tasks on their subordinates in the name of discipline, and also the use of that same authority to send subordinate rivals into unnecessary danger. This chapter updates the term for the Iraq and Afghanistan Wars here through examining wartime memoirs that were based on wartime blogs, and additionally, it argues that chickenshit serves as the master narrative for the wars. By providing a reference point for all of the other narratives to find meaning, it shows the direct connection to the cowardice behind the decision to invade Iraq, how the all-volunteer force assisted that to happen, and how poor leadership and coordination resulted in frustrations and risk aversion all the way down to the tactical level.

"In the Mushy Influences of the Current Times" takes its title from a Walt Whitman quote about how the real horrors of the Civil War would never be written about in a meaningful war. Paul Fussell extends this idea for World War II, arguing that literature of any kind could not reproduce the horrors of World War II for the American public, and that wartime novels might be the least capable of doing so. These prominent authors are not alone in believing this, but the current literary fiction coming out of the wars in Iraq and Afghanistan is proving them wrong.

Through interviews conducted for the purposes of this book with authors of these wars, two veterans and two civilians, this chapter argues that fiction is in fact capturing the truths of these wars at a far deeper level than the preceding nonfiction works of the same subjects could. These authors argue that literary fiction allows them to write about wartime experiences from all perspectives and with greater empathy, it allows them to examine emotional truths of the wars, and it allows them to broaden the cultural conversation about the wars and include the American public in those conversations. Through literary fiction, they are also able to challenge the existing narratives surrounding the wars, and to break through the political and cultural suppositions that have prevented a real conversation on these wars and their costs to the nation.

Although they discuss different narratives, or different aspects of narratives, the chapters of *War Narratives* combine to underscore the deep complexities inherent in wartime decisions and actions, complexities that are often excluded from debates about the wars. *War Narratives* is not intended to offer new truths, but rather to suggest that, when taken alone, these narratives merely allow for a limited understanding of war, its costs, and its effect on the broader society. Because these narratives arose from an unprecedented divide between the civilian public and its military as well as an increasingly mediated information landscape, as argued here, understanding the complexities that challenge these narratives is crucial to ensure that the public is not further manipulated by them in the future.

1

Democratization and Nation Building

The Narrative

JUST NINE DAYS after the September 11 attacks, President George W. Bush addressed a joint session of Congress. He faced a difficult task. He needed to console a grieving nation, to explain the horrific attacks, and to communicate that his government had the nation's security at the top of its agenda.

At one point during his speech, Bush cited the question that he believed many Americans were asking: "Why do they hate us?" Answering his own question about the September 11 attackers, he stated, "They hate our freedoms: our freedom of religion, our freedom of speech, our freedom to vote and assemble and disagree with each other." This rhetorical technique allowed him to make sense of the attacks for the nation, and more important, it allowed him to establish his narrative for the nature of the enemy the nation faced. The enemy of Bush's narrative was one that stood counter to beliefs that were central to American culture and history. And because the attackers were willing to kill Americans for these beliefs, America must be prepared to fight back.

Bush's narrative, that radical Islamic fundamentalists hated America's freedoms, became central to his justification for the war in Afghanistan, and later, the invasion of Iraq. Because it framed the enemy as dangerous and backward, it also set the conditions for his commitment to democratization and nation building in both countries. The challenge of relying on such an ideological justification proved to be that the Islamic extremists who supported the attacks had a different narrative, one that resonated with existing master narratives within the Islamic world, and one that would allow the extremists to pursue their own forms of democratization and nation building in the years that followed.

The driving storyline of extremist Islamic rhetoric, often referred to simply as "the Narrative," states that the West and its allies are continuing in a historical effort to destroy Islam. More particularly, the Narrative states that the United States, Israel, and puppet Islamic leaders, are continuing the historic efforts of outsiders in conspiring to eradicate, exploit, and humiliate Muslims. That some westerners view this as proof that jihadists are freedom fighters and that others bristle at this characterization does not matter; variations on the Narrative are some of the most powerful tools wielded against the West by extremist Islam, providing a common ideological reference point, powerful recruitment tools, and justification for extremists' violence.

A cursory examination of the literature of extremist Islam reveals overt references to the Narrative. In his 1996 manifesto, *A Terrorist's Call to Global Jihad*, Abu Musa'ab al-Suri states, "the modern Crusader-Jewish, American-led campaign against the Arab and Islamic world has clearly announced its goals: total elimination of the civilizational, religious, political, economic, social, and cultural existence of Muslims." In his 1996 fatwa, as translated in *Princeton Readings in Islamist Thought*, "Declaration of War against the American Occupying the Land of the Two Holy Places," Osama bin Laden writes: "It is no secret that the people of Islam have suffered from aggression, iniquity, and injustice imposed on them by the Zionist-Crusaders alliance and their collaborators, to the point where Muslim blood has become the cheapest and their wealth as loot in the hands of the enemies."

But it is not only in the writings of the major jihadist leaders that you find the Narrative. By invoking it through mass communication, the Narrative has come to permeate parts of the broader Islamic worldview. It is common in the media communications of extremist Islam and even in their abundant rumors.

Antiterror leaders in the West have long noted the importance of understanding and combating the Narrative in postmodern warfare. After the rise of the Islamic State of Iraq and Syria, or ISIS, in 2015, President Obama addressed the topic directly, saying "that narrative becomes the foundation upon which terrorists build their ideology and by which they try to justify their violence," as he outlined a plan to defeat the Narrative through a campaign that includes counterterrorism operations, social media, and encouraging Islamic clerics to weigh in. Shaykh Abdullah bin Bayyan, a prominent but controversial Islamic scholar, was among the first to issue a fatwa against ISIS in late 2014, which some have seen as a step toward an effort to weaken the Narrative over time.

Scholars and strategists have examined the sources of the Narrative's power for years, but to limited measurable results. Extreme Islam remains active and aggressive; the reform movement within Islam remains unpersuasive; and the United States and its allies remain as much on the brink of war as they were in 2001 before they invaded two Islamic nations bent on democratization.

In his essay in the *Atlantic*, "What ISIS Really Wants," Graeme Wood asserts that two key misunderstandings of the Islamic State are the reasons for the West's failures. The first error, Wood writes, is that "we tend to see jihadism as monolithic" when it really consists of many factions. Second, and specific to ISIS, Wood asserts that westerners are misled by a "well-intentioned but dishonest campaign to deny the Islamic State's medieval religious nature."

Wood's assertion that extremist Islamic groups have diverse ideologies and strategic goals is a crucial first step. Although his focus is on ISIS, he alludes to a major difference between ISIS and Al-Qaeda: where ISIS wishes to establish a caliphate based on original Islam, Al-Qaeda has focused on a more amorphous goal of decentralized terrorism prosecuted through a global network. Stated differently, ISIS aims to establish a premodern state while Al-Qaeda aims toward a postmodern movement. Both rely on narratives to achieve their goals and to erode the modern political and religious barriers, respectively, they confront.

Scholars' efforts to understand and combat extreme Islamic narratives have yet to provide a full picture. The shortcomings in understanding narratives are precisely for the reasons Wood describes: they tend to view jihadism as monolithic, namely, that all jihadist organizations desire a geographic caliphate, and they fail to recognize that the extraordinary differences in strategic outcomes are driven by extremely different philosophical worldviews. In doing so, these scholars have attempted to understand the narrative in modern or postmodern terms instead of applying those lenses to the various philosophical foundations of extremist groups themselves.

In the 2012 book *Narrative Landmines: Rumors, Islamist Extremism, and the Struggle for Strategic Influence*, Daniel Bernardi and his coauthors examine

Islamic narratives through their most basic element, namely, rumors, while Jeffry Halverson and his coauthors study the hierarchy of narratives in their 2013 book, *Master Narratives of Islamist Extremism*. Both books largely agree that narratives operate rationally within formal frameworks and that they are used to achieve the jihadist goal of establishing a global caliphate. However, they disagree on nearly all of the particulars, from whether narratives can be strategic tools to the appropriate use of modern or postmodern theories when understanding them.

Narrative Landmines studies how narratives manifest locally through rumor, and how rumor drives broader beliefs. For them, "rumor is a shorthand term for speculation, half-truths, and misinformation in the form of stories that, to some groups, appear to offer rational cause-and-effect explanations of events." Previous scholars argued that rumors are nonnarrative in nature, an idea that these authors contest, arguing instead that rumors "are, in essence, local stories that participate in larger narrative systems that, given the right mix of cultural references and ideological formations, end up volatile when placed in the service of either an insurgent or counterinsurgent agenda." To them, rumors are actually "narrative IEDs," mimicking the cheap and effective tactical weapons insurgents use in Iraq and Afghanistan.

If rumors are tactical weapons, then they are most effective when they complement narratives used as strategic communications. In a broad sense, a narrative is "a system of interrelated stories that share common elements and a rhetorical desire to resolve a conflict by structuring audience expectations and interpretations." For their study of rumor in wartime, they define the concept as "a continuous process of understanding in which external data are constantly ordered and re-ordered into patterns that make sense to people." A "narrative landscape," then, is "a system of narratives" that refers "to the complex array of narratives that circulate within a specific social, economic, political, and mediated environment—the narrative context that gives rise to rumors and in which rumors compete for prominence."

Unlike rumors, narratives comprise stories, which they argue, "consist of sequences of events involving characters in setting and are recounted for rhetorical purposes." There are three components to stories that make them meaningful and understandable to audiences: archetypes, story forms, and master narratives. These components are crucial for successful narratives as they are derived from "larger cultural systems."

In narratives, archetypes represent easily identifiable shorthand for versions of people, places, and events the Muslim world has interacted with throughout its history. They range from tyrants and hypocrites, to crusaders and colonizers, martyrs and champions, and include the oppressive pharaohs of Egypt, to the Mongolian military leader Hulagu Khan, who sacked Baghdad in the thirteenth century, and others. Archetypes act through story forms, or "recognizable patterns of storytelling that are easily comprehended because of their familiarity." Common story forms include Invasion, Noble Sacrifice, Reward, and Deliverance, which are intended to persuade the audience that they are an aggressed underdog, and dying and even killing other believers for their cause is just. Master narratives move "across historical boundaries, resolving archetypal conflicts through established and literary and historical forms," and include the Crusades and colonization, which depict westerners invading Muslim territories to "exert their political power to extract natural resources (Islamic wealth and property) from the rightful Muslim owners."

Master Narratives of Islamist Extremism provides an alternative understanding of narratives. Its proposed structure operates through "vertical integration" between three levels of narratives, with communication at the lowest level being more "concrete" and "manifest" and moving toward the more "abstract" and "latent" at the highest level. Vertical integration of narratives allows Islamic extremists to achieve their primary goals, which the authors refer to as resistance, rebuking, and renewal.

The rumors described in *Narrative Landmines* are most closely aligned with "personal narratives," or "systems of stories experienced, remembered, and told by individuals." Personal narratives integrate with "master narratives," which the authors describe as "narratives whose component stories, by virtue of being widely shared and repeated across time, have become deeply embedded in a particular culture." The majority of their work is a detailed analysis of thirteen master narratives that they have identified in extremist literature, including the crusader and hypocrites narratives.

Contrary to the central argument of *Narrative Landmines*, Halverson and his coauthors argue that master narratives are not strategic. Instead, master narratives combine "at the highest level of abstraction" to form a "rhetorical vision," which is similar to the "narrative landscape" defined in *Narrative Landmines*. A rhetorical vision uses story forms and archetypes to integrate "master narratives in an emotional and logical sense," providing common "resources for constructing personal narratives and developing goals for actions and behaviors."

While both sets of authors base their arguments on highly structured narrative models, they have opposite views when it comes to addressing modernism and postmodernism as they apply to Islamic narratives. The authors of *Narrative Landmines* see modern rigidity as the fundamental source of the West's inability to combat the Narrative. Conversely, the authors of *Master Narratives of Extremist Islam* favor a decidedly modern approach.

The authors of *Narrative Landmines* provide an important analysis of the concept of objective truth in wartime rumors. "The standard in rumor studies has been for many years to define, and thus study, a 'rumor' as information or belief about a situation that is of dubious veracity, a statement that does not meet 'secured standards of evidence,'" they write. "Although dubious veracity and unclear origins are central characteristics of rumors, an overemphasis on veracity and truth claims isolates rumors categorically, rather than recognizing their impact and dynamic functions, especially with regard to strategic communications contexts."

Instead, these authors call for rumors to "be understood in their cultural contexts." This means "the truth of a situation presented to any given individual has more to do with how that individual organizes available data than any objective and idealized 'truth.'" It is the primary function of narratives, therefore, "to make sense of that body of data," which narratives do by offering "a means of uniting culturally provided templates to include story forms and archetypes with data such as stories, rumors, histories, and the like." In the end, "'truth' becomes less about facts and evidence and more about coherence with pre-existing and prevailing understandings." This is a postmodern understanding of rumor and narrative, making them a highly subjective force that actively deconstructs modern structures and beliefs.

Master Narratives of Extremist Islam builds its case for a modern understanding of narrative on a modern understanding of Islam. In modern Islam, the authors argue, being able to "quote extensively from the Qu'ran by memory creates an aura of knowledge and piety in a speaker and establishes a hierarchical position that situates the speaker apart from (or above) the audience on the basis of his or her knowledge of the sacred text, infusing or encoding his or her words with transcendent authority." Modern Islam allows those in higher positions to dispense doctrine and adjudicate legal matters in accordance with the various sacred texts of Islam.

Islam's master narratives, then, "are thoroughly 'modern' (in a macro sense) and their ideological uses are definitively 'modernist.'" The authors concede that "some readers may ask why [they] are relying on elements of modernist notions of narrative" when so many in their field have "moved beyond modernist conceptions like 'master narratives'" to understand the way narratives work. They respond that "while postmodern theories may well represent attractive resources for how we might learn to *counter* extremist arguments (in the same ways that postmodern theories provide rich textual resources to counter modern conceptions of narratives) it is vital to remember that Islamist extremists emphasize the articulation of 'transcendent' values derived from a 'grand' or authoritative text."

These distinctions highlight the shortcomings in both of these works, which are the same shortcomings proposed by Wood. First, they view Islamic terror organizations as monoliths, at the very least implying that they all seek the establishment of an Islamic caliphate. Second, central to both works are misapplications of the modern and postmodern lenses to the narratives themselves. Narratives should simply be seen as tools used to achieve the different strategic goals of organizations like ISIS and Al-Qaeda. Because modernity and postmodernity describe philosophical foundations, these lenses are better applied to understanding the organizations and to their movements.

In an article titled "A Tale of Two Jihads: Comparing the Al-Qaeda and ISIS Narratives," Naureen Chowdhury Fink and Benjamin Sugg suggest that this monolithic understanding of extremist Islamic organizations is understandable due to their many surface similarities. Through their examination of the official publications of Al-Qaeda and ISIS, *Inspire* and *Dabiq*, respectively, they note that both organizations demand jihad from all Muslims and that they rely on the broad use of the Narrative to justify their actions. However, beyond those surface similarities, Al-Qaeda and ISIS have vastly different "ultimate aims," which are reflected by the narratives they generate.

The authors identify that the driving narratives found in Al-Qaeda's *Inspire* revolve around three recurring themes. First, the magazine glorifies "militant activities and terrorism"; second, *Inspire* provides a "DIY guide for weaponry"; and third, it encourages "'lone wolf' acts of terror by individuals that can have absolutely no formal association with Al-Qaeda." All of the elements of Al-Qaeda's narrative appear to empower independent, uncoordinated action toward achieving Bin Laden's one-time goal of a "defensive jihad, a battle to promote and protect the *Ummah*, or global Muslim community, from the onslaught from the West."

Unlike Al-Qaeda's narrative, the authors show that ISIS's narrative revolves around a centralized, organized plan for the establishment of a formal Islamic State. *Dabiq* calls for support of the caliphate through appealing "to doctors, engineers, and professionals to make hijra (migration) in order to assist the construction of an Islamic government," the authors write. It also advances the narrative of political Islam by highlighting the type and level of "social services, defense, and dignity" that Muslims will be able to enjoy once protected by the newly established caliphate.

These strategic differences between ISIS and Al-Qaeda complement their philosophical differences. ISIS is driven by a premodern worldview toward the establishment of a premodern, medieval caliphate, while Al-Qaeda is driven by a postmodern worldview toward the facilitation of perpetual global jihad. Narratives play a key role within these strategic and philosophical frameworks, with ISIS using narratives to replace the constraints of modern political constructs, and Al-Qaeda using narratives to replace the constructs of modern Islamic jurisprudence and doctrine to justify their violent actions as holy and pious.

If ISIS's strategic objective is the establishment of a premodern caliphate, then it is akin to movements in political Islam, the greatest distinction being ISIS's reliance on a premodern worldview. Regimes within Islamic states have long used official or national narratives to rally their people in support of their vision for the future, to rise up against continued enemies and challenges, or both. In her book *Official Stories: Politics and National Narratives in Egypt and Algeria*, Laurie Brand studies the creation, evolution, and circulation of these narratives, and in a subsequent essay, she argues that ISIS has already developed similar national narratives in support of its strategic objective of forming a caliphate.

According to Brand, national narratives are the nation's story as it is told and understood by the people. They amount to the nation's collective social, ethnic, religious, and cultural identity, especially as they relate to the nation's understanding of its story through historic trials, victories, enemies, and heroes. They establish legitimacy to rule for the regime, and once implemented, regimes can select and shape elements of their national narrative to adapt to the current challenges the regime faces.

"No political leader or elite, even in authoritarian states, rules solely through the threat of coercive violence," Brand writes. Additionally, they require "some level of support, or at least acquiescence, from the people over whom they rule." This support is often developed not entirely through formal levers of power like the military, the courts, and other institutions, but through "less tangible elements" organized in a "complex configuration of values, customs, political principles, and social relations accepted throughout society and its institutions at a given historical moment."

Attempts by regimes to control these variables have often come from the efforts to control public discourse, as Brand notes, especially as they attempt to establish official narratives. They can be seen in a regime's investment in cultural productions, such as plays and literature, through asserting control over educational texts, through speeches, policy programs, charter documents, or other official statements. Brand places the use of official narratives by states or state actors within the literature of propaganda, which historically has been effectively used to shape public opinion in times of crisis.

Establishing a national identity and unity are major goals of national narratives, but Brand argues that they are not desired ends. Regimes use narratives to establish their legitimacy, which she describes as "the process by which a leadership secures the people's acknowledgement of its right to rule." This process is accomplished through using narrative by obscuring their "state-building efforts intended to serve the citizenry in general," on one hand, and policies intended to sustain and reinforce the regime's hold on power, on the other. Without linking the interests of the people with the actions in support of sustaining the regime's power, Brand argues, "a key part of the narrative's legitimizing project will fail."

If such a linkage is created through national narratives, then the regime is able to rely on the Narrative to rally support for its vision for governance. That is, by establishing itself as legitimate leaders, the regime can further legitimize its "guiding ideology and the policies that flow from it." By fitting their current existential threats or challenges into their established national narratives, regime leaders can further use narratives to provide the populace with a rationale for their personal responsibility to their nation, and "exhorting the people to serve, protect, and even die for them."

While *Official Stories* is exclusively about the use of national narratives by Egypt and Algeria in the twentieth century, it also provides a remarkable overview of how and why Islamic states have circulated official narratives to establish their legitimacy and build their nations. Because ISIS is a derivative of political Islam, Brand's argument regarding the use of national narratives found in *Official Stories* can be applied as well. Brand does exactly this in a piece in the *Washington Post*, "The Islamic State and the Politics of Official Narratives," written only a few months after the establishment of the Islamic State was declared in 2014.

Brand argues here that the earliest actions of ISIS suggest that they are poised to develop a national narrative to legitimize their own government and build a case for national unity around the concept of a premodern caliphate. Brand points out that ISIS established "its first major seat of power" in Mosul, Iraq, where "the basic institutions of its nascent state can be most effectively observed." These institutions consist broadly of the basic civil services aligned with the tenets of Islam: prisons and courts, mechanisms for complaints and tax collection, and even "morality patrols" intended to ensure that the populace lives up to sharia law.

In addition to these formal efforts, Brand also notes an early move by ISIS to establish the criteria for "legitimate education." Examining an August 2014 release from the Islamic State's Bureau of Curriculum titled "General Directive to All Educational Institutions," Brand analyzes ISIS's efforts to establish the initial model for educational instruction under the Islamic State. According to the directive's first order, schools should immediately remove from their curriculum all training in humanities subjects, such as music, art, civics, social studies, and history, but also in mathematics and religious studies, to include Christian and Islamic education. Brand notes that the directive allows for these subjects to be replaced, which she concludes to mean "what is underway is not the wholesale abolition of most courses of study, but rather the first stage in a massive reworking of the curriculum."

This reworked curriculum will establish a new national narrative for the Islamic State. This new narrative will "'correct' the history and mission promulgated by the former colonial power" and in its place, "an heroic story would be constructed, aimed at building a unified national identity, establishing the vision of that nation, and—crucially—consolidating power through reinforcing the regime's legitimacy to rule." Specifically, the directive calls for the reworked curriculum to replace national symbols and identities with Islamic ones, to remove photographs that defy sharia, and to replace references to evolution with attributing all acts of creation to Allah.

These early actions, especially when combined with the Islamic State's expressions of violence, can fairly easily be seen as efforts to legitimize a conservative, authoritarian, premodern state. However, they should also be seen as efforts to delegitimize and erode support for competing modern political structures and practices through actions that directly support their narrative. In addition to those just cited, the Islamic State has reportedly commandeered oil production facilities in order to sell the commodity and fund their efforts; they appear to continually disregard global treaties and practices by exacting a ransom for foreign hostages, and even murdering some on video for the largest response; and, Brand points out, ISIS has destroyed the border crossing between Syria and Iraq, "which the Islamic State framed in terms of overturning the legacy of Sykes-Picot, demonstrated its ideological rejection of existing state boundaries as it extended its own realm." Unlike other Islamic nations determined to gain legitimacy through establishing and using national narratives, inherent within the ISIS narrative and actions is the message that it cannot exist alongside competing practices, beliefs, or even states.

Unlike ISIS, Al-Qaeda's objective is maintaining a disruptive global jihadist movement. In order to accomplish this, Al-Qaeda has applied a form of postmodern thought to Islam, deconstructing Islam's authority structure and replacing it with a more subjective and democratized form. Although it is not a formal study of narratives, Al-Qaeda's postmodern form of Islam is explored in great detail in Faisal Devji's book, *Landscapes of the Jihad: Militancy, Morality, Modernity.*

Landscapes of the Jihad draws a significant distinction between political Islam and global jihad, though because it was written in 2005 it does not consider the rise of ISIS. To Devji, political Islamists aim to create a geographic caliphate, while global jihadists aim to create a metaphysical one. Within the deconstructed Islam of global jihadist movements like Al-Qaeda, narratives have replaced traditional Islamic doctrine as the central authority of Islam, and are also used to justify their tactics.

Although Devji never uses the term *postmodern,* his argument suggests that he is referring to a deconstructed Islam created through efforts by global jihadists to "wrest the jihad away from the juridical language of the state and make it a strictly individual duty that is more ethical than political in nature." Global jihadists have focused on the "destruction of the traditional forms and distinctions of Islamic authority," which has resulted in "the democratization of such authority among all manner of groups as well as individuals." Decrees and decisions once dictated by Islamic doctrinal and juridical authorities are now "given over to individual rather than collective examination."

The fragmentation of traditional Islam influences the structure and language of the movement as well. Instead of Islamic groups committed to political change, the jihad has instead provided an umbrella cause and organizational structure for various groups and various interests, creating an "extraordinarily diverse membership, one that is not united by way of any cultic or ideological commonality, to say nothing about that of class, ethnicity or personal background." For Devji, global jihad operates almost exclusively as a loosely affiliated network, diminishing the centrality of the Arabic culture and language, and has even replaced traditional Islamic architectural references, like Islam's five pillars, with Sufi imagery.

As with all forms of postmodern philosophy, this entails rejecting claims to objective truth and claiming instead a necessary tolerance for the individual or subjective truth experienced by others. Devji addresses this directly, claiming, "militant Sunnism has abandoned the theological disputes of the past, based as these were on differing claims to the truth, and adopted a democratic narrative of enmity instead." By this he means "it is no longer arguments about truth that animate such militants, only a desire for the recognition and respect of their neighbors, who are accused of insulting their sanctities while at the same time claiming to be fellow Muslims."

Although Devji does not go in great detail into what he means by jihadists adopting a "democratic narrative of enmity," his assertion here is at the center of his disagreement with the authors of *Master Narratives of Islamist Extremism* and *Narrative Landmines.* To these authors, narratives are defined as important communications tools used to drive their fellow believers toward an Islamic state. But to the global jihadists, narratives have replaced Islam's fundamental objective truths. They are used to explain the world in terms that are aligned with the Narrative, but that are interpreted subjectively by the individual jihadist.

This subjective interpretation is available because a belief in mystical engagement with the divine allows leaders of the global jihad to interface directly with God. "Everything we know about Al-Qaeda as a religious movement compares favorably with Sufi or mystical brotherhoods," Devji writes, "even if these happen to be disapproved of by members of the movement itself." This results in a significant reliance on the occult, on dreams as direct contact with God, and a view of Al-Qaeda's leaders as a type of prophet archetype, "as spiritual authorities endowed with a grace that is unrelated to bookish learning of the kind that would mark a traditional cleric or even a fundamentalist." This not only provides an easy way to circumvent the traditional authorities of Islam, but it is also an enhancement to a democratized Islam in that it is used to reinforce the personal and ethical expectations of postmodern Islam.

Deconstructed Islam coincided perfectly with the new media age, Devji argues, with the internet allowing beginners to have the same voice as established authorities. The new media age also assists in making the jihad available globally and impossible to ignore. Devji argues that, for most, the jihad "is experienced visually, as a landscape initially made available by way of the international media and then redacted in conversations, posters, literature, art-work" and so on. Acts of jihad are the primary visual associated with the religion, and through these depictions, "the jihad's battlefields become sites of a global Islam." This media focus on Islam's violence not only drives jihadist attention and resources, but it also creates a new standard of personal and ethical Islam and unites "Muslims and non-Muslims alike in a common visual practice."

Devji argues that this media dynamic within global jihad creates a universal, metaphysical caliphate, which is the ultimate goal for global jihadists. Through media representations of jihad, a "generic Muslim" is created, "one who loses all cultural and historical particularity by his or her destruction as an act of martyrdom." This allows the jihad to achieve universality, allowing all viewers, even future martyrs, to experience the jihad in ways that are "entirely abstract and individual," allowing them "to break with locally available forms of Islamic authority." It also forces all who view martyrdom as "an ethical choice to support either its Muslim victims or their infidel oppressors—and all who make this choice are held responsible for it, having become participants in the jihad irrespective of their knowledge about its truth." That is, it is not the truth behind the cause of global jihad that requires its viewers to make this choice, but the "spectacle of martyrdom" itself.

It is this notion of truth within global jihad that clearly defines it as a postmodern movement, one that is made especially possible because of these new media characteristics. "Media images of martyrdom have no epistemological status," Devji writes, because truth is subjective and is never "a subject for discussion in the holy war." Truth, then, in true postmodern form, fits a similar definition for narratives provided in *Narrative Landmines*: truth is subjective, self-referential, and ultimately as flexible as it needs to be to justify Islamic fundamentalists' tactics and to pursue their strategic goals.

In his 2017 book, *Words Are Weapons: Inside ISIS's Rhetoric of Terror*, Philippe-Joseph Salazar provides a sobering update to radical Islamic groups' use of rhetoric and narratives. In his book, the French scholar of rhetoric describes how ISIS has effectively used rhetorical tools to achieve its goals, and he also points out how ineffective Western counternarrative efforts have been. Further, he suggests that ISIS may have already embraced the innovation of a democratized Islam as described by Devji, and incorporated it into their political Islamic nation-building campaign, resulting in a deadly ideological merger.

In *Words Are Weapons*, Salazar contrasts the rhetoric of ISIS with that of the West in order to offer his recommended approach. He states that when westerners hear the rhetoric of radical Islam, like that of the Narrative, it sounds "florid, exaggerated, poetic, and ostentatious." However, to their ears, American political rhetoric, like Bush's "they hate us for our freedoms" narrative, sounds "sterile, rhetorically bland, and poetically deficient." In order to defeat their rhetoric, he argues, the West must understand it. Once it is understood, those in the West must name their actions, and do so in terms that resonate in negative terms with their cultural narratives.

Perhaps more compelling is Salazar's conclusion that ISIS too has embraced a postmodern jihad. He describes ISIS as a populist movement bent on removing elites from the existing hierarchy, a movement that embraces cultural relativism as a "useful tool" for accomplishing its objectives, and, as Devji describes of Al-Qaeda, using shocking acts on video to personalize the war for everyone. ISIS represents, in other words, a deadly new mutation of radical Islam that has learned lessons from both its efforts in modern political Islam and Al-Qaeda's postmodern global jihad.

2

The Deciders

The War of Choice Narrative

MANY accusations have been leveled against President George W. Bush and his administration regarding the decision to invade Iraq after the attacks of September 11. Some charge that the president cherry-picked evidence, some charge that he was acting on behalf of business interests, and still others associate his decision to invade with arrogant foreign policy. All of these charges align with a narrative that asserts that President Bush engaged in a war of choice in Iraq.

The war of choice narrative states that the president and his cabinet were not forced or drawn into war in Iraq, but rather, they chose it. It implies that the president made his decision to take the nation to war on his own, he made his decision to do so early after the attacks of September 11, 2001, and that once he made up his mind, he lacked the curiosity or humility necessary to change course. Conveniently, the war of choice narrative also allows those who embrace it to assign blame to the president for every aspect of the calamitous war in Iraq that followed his decision made in isolation.

In political rhetoric surrounding the war in Iraq, the charges of the war of choice narrative are compelling and powerful. The narrative allowed its proponents to claim that while the invasion of Afghanistan was righteous and warranted, the decision to invade Iraq was wrong and optional. Some have infused their rhetoric against the invasion with the subtle accusation that a war of choice is actually a war of aggression, which is a war crime.

This narrative is a generous political gift to the Bush administration and its opponents, but it does not properly reflect the true number of people and entities that were actually involved in the decision to invade Iraq. Instead, it changes the debate into political terms, which inherently allows everyone involved to position themselves for maximum credit and minimum culpability. When the Iraq War is discussed in these stark terms, important topics like presidential war powers, congressional oversight, and the all-volunteer military go unconsidered, but considering these topics reveals that it was not merely a presidential choice that led the nation into such a war, but one that was chosen by many, and especially those who had volunteered for military service.

The two sides of the war of choice narrative were explored by sub-cabinet-level officials during the Bush administration through Richard Haass's *War of Necessity, War of Choice* and Douglas Feith's *War and Decision.* Haass, who served on the National Security Council under President George H. W. Bush and as the director of policy planning in the State Department for George W. Bush, contrasts the first President Bush's decision to invade Iraq with the second President Bush's decision. Feith, the undersecretary of defense for policy in the second Bush administration from 2001 to 2005, provides his administration's account of the decisions and actions that led to the Global War on Terror.

Haass, who resigned in protest from the second Bush administration, defines a war of necessity as one that is prosecuted to protect or advance vital national interests, which he argues fits the justification for the first Iraq War. Conversely, wars of choice are defined as wars in pursuit of nonvital interests or when other options are available, a definition he applies to the second Iraq War. "There are wars of necessity and wars of choice," he writes in his introduction. "Confusing the two runs the danger of ill-advised decisions to go to war."

Haass, like many proponents of the war of choice narrative, asserts that George W. Bush had chosen to go to war in Iraq well ahead of the invasion, and in doing so, he argues, the invasion of Iraq was a war of choice twice over. First, because it did not pass the basic national interest test, and second because of the way it was executed. "The relatively low number of troops brought to the theater," he writes, "all but guaranteed the United States and its few coalition partners would not be in a position to assert and maintain order once the formal battles were concluded."

Feith spends a considerable portion of *War and Decision* providing the opposite perspective. He suggests that while the military was directed to start developing military options for Iraq as early as September 2001, the decision to invade was based on a deliberate process. Regarding the final decision, he writes that George W. Bush weighed "the risks of war against the risks of leaving Saddam in power" before deciding that "it was unreasonably risky to allow Saddam to choose the time and place for turning Iraq's ongoing low-level confrontation with the United States into a high-level conflict." It was not, in Feith's telling, a brash step toward war.

Feith is careful to show the Bush administration's thoughtful approach to the invasion, and he also shows how others participated in that decision. Outlining the same case that many other Bush administration officials would later make, Feith argues that war was first declared on the United States through the terrorist attacks on September 11, 2001; that Saddam Hussein had ample time and opportunity to choose to avoid the conflict; and that the United Nations also forced the decision through their failed actions. Citing the previous failures of the international community—"diplomatic protests, Security Council resolutions, weapons inspections, economic sanctions, no-fly zones, no-drive zones, and limited military strikes"—he notes that the president was left with few options. Regarding the decision to invade, Feith concludes, "reasonable people differed then, and differ now, on whether war was the right choice."

While Haass and Feith argue the opposing positions of the war of choice narrative, key leaders of the Bush administration provide additional details from their perspectives through their memoirs from that period, President Bush's *Decision Points*, Vice President Cheney's *In My Time*, Secretary of State Rice's *No Higher Honor*, and Secretary of Defense Rumsfeld's *Known and Unknown*. These works parse the war of choice narrative in order to denounce its unfavorable aspects while embracing its favorable aspects, a significant gamble, but one that allows them to look like visionaries if their actions eventually led to a revolutionary transformation of the Middle East or like courageous actors responded during a difficult time if their efforts eventually failed. Cheney sums up this position perfectly when he writes, "when historians look back, George Bush's decision will stand out as one that made a difference for millions and put history on a better track."

Even with quotes like this, it is clear in the works of these administration officials that they see the central charge of the war of choice narrative as potentially damning to their legacies, and at times they address it directly. Rumsfeld writes bluntly, "the unilateralist charge against Bush was a preposterous charge." Rice describes how the narrative worked in the psychology of the administration's opponents, writing, "since many people believed that we'd already decided to go to war against Iraq, sinister interpretations suggested that we were preparing to use military force against all three states [of the axis of evil]," which amounted to, in their minds, a preemptive declaration of war on North Korea, Iran, and Iraq.

To support their rejection of claims regarding the president's unilateral choice for war, these administration officials describe the number of key international players who also had a part in the decision. President Bush writes that Hussein and Iraq had been given numerous chances, "it had been more than a decade since the Gulf War resolutions had demanded that Saddam disarm, over four years since he had kicked out the weapons inspectors, six months since I had issued my ultimatum to the UN, four months since Resolution 1441 had given Saddam his 'final opportunity,' and three months past the deadline to fully disclose his WMD." In a section of her book that details the actions of the United Nations that preceded the second Bush presidency, Rice titles a chapter of her book, "Confronting the International Community with a Choice," where she provides a litany of previous failures by the United Nations to enforce threatening resolutions against Hussein.

President Bush assigns additional blame even further than his global counterparts, identifying a culpable and unsympathetic antagonist, Saddam Hussein, who "chose war" through his actions during the diplomatic efforts that preceded the war. Like Feith, these authors also point out that Al-Qaeda had also participated in the choice for war when they declared war on the United States through their actions on September 11, 2001. Combined, their point is clear: President Bush did not choose to go to war; Europe's inaction and weakness, Hussein's disrespect for the global community, and the aggressive attacks of global jihadists did.

Part of the reason that the war of choice narrative is compelling, though, is that Bush's image and actions seemed to support it. Rice concedes that there is room for the argument Haass makes about the two Bush presidencies and their wars: "the policies of the two men would be compared and contrasted: realism versus idealism; diplomacy versus confrontation; compromise versus absolutism; prudence versus plunging." She rejects these simplistic characterizations, adding, "yes there were differences in style and temperament, with George W. Bush quicker to anger and less given to shades of gray."

This nature of being "quicker to anger and less given to shades of gray" became a crucial part of the George W. Bush identity, especially to those driving the war of choice narrative. But even with the image of the president as a man of action, the actual decision to go to war was played out in a process that could only be described as in the gray area, much like the administration's efforts to embrace the positive and reject the negative aspects of the war of choice narrative. In choosing their approach for dealing with Hussein, the administration settled on "coercive diplomacy," an approach that sought the best aspects of both military and diplomatic solutions throughout, while driving toward their preconceived outcome.

Rice, writing about her time as national security advisor, takes credit for telling the president about coercive diplomacy, which the president describes as a two-pronged approach. "One was to rally a coalition of nations to make clear that Saddam's defiance of his international obligations was unacceptable," he writes. "The other was to develop a credible military option that could be used if he failed to comply." While the two tracks would remain parallel, at least initially, eventually they would converge, with the decision point for war occurring immediately before convergence. Important to their view of the war of choice narrative, Bush adds that this course of action would ensure that the decision to pursue diplomacy or military options would be Saddam Hussein's alone.

There are obvious challenges to this approach. First, it is easy to see how the decision to pursue a policy that merges military force with diplomatic pressures would almost undoubtedly lead to war with Hussein's Iraq. Such would be a high-stakes gamble with traditional actors, and with Hussein's track record over the recent decades, coercive diplomacy would be an incredibly aggressive strategy. Second, it is difficult to believe that the nation's decision to go to war could really be assigned to a foreign dictator.

Coercive diplomacy began in 2002, after Bush had achieved initial success in Afghanistan. The United Nations weapons inspectors were the primary tool for this approach, yet their efforts proved fruitless against Hussein's stonewalling and stalling. Cheney and Rumsfeld, hardened by their experiences leading up to and following the first Gulf War, were convinced that Hussein viewed America as weak and pushed harder for the military side of the equation, while Rice and Bush portray themselves as being intent on letting the diplomatic process fail before choosing to introduce military force. In the end, this dual strategy ensured a march to war.

But even if war were heavily favored in this approach, the United States Constitution provided a significant barrier to a unilateral presidential decision to invade Iraq. The Constitution requires congressional authorization for war, but it also provides for powers to the president as the commander in chief, a duality that provides these authors an argument against the war of choice narrative by allowing them to show the complicity of Congress in Bush's decision to invade Iraq while also asserting his legitimate authority to make the decision alone. Although these arguments may seem contradictory, they align with an approach aimed at rejecting the negative aspects of the narrative while embracing the positive.

As with most aspects of the war of choice narrative, there are dueling perspectives on questions of presidential and congressional wartime authority, and Haass and Feith provide helpful insights. Haass describes both sides of the narrative by noting that "no president wanted to have his hands tied, and many in Congress did not want to give the president all the powers that accrue to him if war is formally declared." Feith expands on Haass's comment by arguing that Congress did in fact vote to authorize the war in legislation in October 2002 when it authorized the president to use the military to "defend the national security of the United States against the continuing threat posed by Iraq," and "enforce all relevant United Nations Security Council resolutions regarding Iraq." Interpreting this congressional authorization would provide the greatest weakness to the war of choice narrative and the greatest strength to the arguments made by senior Bush administration officials, who could claim they had acted in full support of the Constitution, the Congress, and the American people.

Feith's seniors in the Bush administration go to great lengths in their memoirs to detail their beliefs on presidential war powers. This is especially true of Vice President Cheney and Secretary Rumsfeld, both of whom had served in Congress before serving in key roles in the executive branch under several presidents. Their experiences deeply shaded their views on the supremacy of presidential war powers over congressional war authorization, which they saw as almost a formality.

Rumsfeld's views on executive war powers evolved over time, starting in his college years. While an undergraduate at Princeton, his thesis was on the Supreme Court case *Youngstown Sheet & Tube Co. v. Sawyer*, where the court had ruled that President Truman's seizure of the plants and materials during wartime had been unconstitutional. Rumsfeld recalls, "I argued in my thesis that the Court's decision was 'timely and reassuring,'" and that "the 1952 case would become an important decision about the limits of executive power in wartime."

His beliefs in the limits on executive wartime power were further solidified less than a decade later when he was a young congressman from Illinois faced with a vote on the Gulf of Tonkin Resolution. "When I voted for the Gulf of Tonkin Resolution, I did have some concerns about its language—I worried it might be interpreted too broadly by LBJ," Rumsfeld writes. He had reason to be concerned, as LBJ used it as a categorical justification for his wartime decisions and later used the resolution to blame Congress for the deteriorating situation in Vietnam. That Rumsfeld wrote his thesis in favor of limiting presidential war powers and felt burned by LBJ on the Gulf of Tonkin vote is remarkable considering that later he supported a view that Congress was largely a barrier to be overcome by a strong executive when exercising his exclusive war powers.

Later, when Rumsfeld was recalled by President Reagan to serve as an envoy to the Middle East following the attacks in Lebanon, his opinions on presidential war powers showed continued signs of this evolution. The Reagan administration wanted to pursue a military response to the attacks in Lebanon, but, as Rumsfeld writes, "the administration's strategy faced another major impediment, namely the United States Congress." By the Iraq War, Rumsfeld's evolution was complete, noting that "the President was the one charged with the tough choice to commit U.S. forces," and that it was a "necessarily lonely decision."

Although he does not openly discuss his transition, Rumsfeld offers some insights into his changing opinions. First was a policy shift with the passage of the War Powers Resolution during the Vietnam War, which required that military force be completed with "sixty to ninety days absent the explicit authorization of Congress." Second, Rumsfeld describes that there had been a culture change in Washington since he was last in the Pentagon in the late 1970s, where "the relationship between the executive and legislative branches had evolved from proper congressional oversight to what was becoming legislative micromanagement."

Like the others, Rumsfeld outlines all of the authorizations that the Bush administration embraced to justify the invasion. Going back to 1991, there had been "seventeen UN resolutions demanding that Saddam comply with various requirements." Rumsfeld sites the Iraq Liberation Act of 1998, which, as he puts it, stated "regime change in Iraq was now the official policy of the United States." He is favorable of Congress's vote authorizing "the use of all 'necessary and appropriate force' in the fight against terrorists," after the September 11, 2001, attacks. And, of course, the resolution of 2002 that had widespread and bipartisan support in Congress, which he portrays as providing definitive authorization for the invasion.

It was the attacks of September 11 that caused the biggest change to his views on war powers, though. After the attacks, he writes, "calls for greater security through prompt action came from every quarter of the country." This was complicated by constitutional design, which intended Congress to be slow and "to promote deliberation and the weeding out of ideas that may be popular for a moment, but imprudent." He notes that "Congress was not intended or organized to meet the demands of operational decision making in a crisis," and that the nation's founders "knew what they were doing when they put the powers to conduct war in the hands of a single commander in chief, not those of a committee composed of the 535 members of the national legislature." His views of presidential war powers had changed dramatically since he wrote his thesis at Princeton, but he also concedes that had they worked closer with Congress on issues related to the war, then perhaps "presidential powers in wartime might as a result be stronger than they are today."

No matter how aggressively Rumsfeld worked to protect the president's wartime authority, his efforts paled in comparison to those of Vice President Cheney. As Rumsfeld puts it, Cheney "appreciated the importance of preserving the President's powers as commander in chief, especially when they were under assault for short-term political reasons." Cheney confirms this in his own words, stating, "much has been written about my advocacy of a strong executive, and it is true that I am a firm believer in protecting the president's prerogatives, especially when it comes to the conduct of national security policy."

Like Rumsfeld, Cheney's perspective on presidential war powers ahead of the Iraq War developed through his long career at the highest levels of government service, but unlike Rumsfeld, he offers no hint that his perspective on the subject might have evolved over time. About his time in Congress in the 1970s and 1980s, he writes that he thought it was "crucial to defend the presidency itself against congressional attempts to encroach on its power." In an appearance on *Meet the Press* around the time of the first Gulf War, he relates that he told the host that he loved the Congress, but he also had considered its limitations when determining whether or not to pursue congressional authorization for the invasion. But if his views on presidential war powers remained steady, his memoir shows that they were also complicated, allowing him to argue that congressional authorization was unnecessary, while also going to great lengths to point out all of the authorizations that the president had received for the second Iraq War.

President Bush and Secretary Rice had far less to say on the subject of presidential war powers, with Rice merely noting that the president alone is responsible for the decision to send troops into combat. President Bush goes slightly further, noting that several of his decisions were well within his role as commander in chief as outlined in the Constitution: setting the vision, determining a consistent doctrinal approach for the nation's defense, and even selling the war to Congress and the American people. His preferred doctrine, of course, like his choice to pursue coercive diplomacy, favored military confrontation. Beyond the general provisions within the Constitution, and the particular power assigned to the commander in chief in Article II, Bush also relied on the congressional war resolution passed following the attacks of September 11, 2001, which provided authorization for retaliation against targets associated with the September 11 attacks. Finally, he cites the Iraq War authorization of 2002.

Regarding this final authorization, President Bush addresses a key element of the war of choice narrative. He writes, "some members of Congress would later claim they were not voting to authorize war but only to continue diplomacy." To point out the flaw in this maneuver, Bush cites the resolution directly, showing that it authorized him as commander in chief "to use the Armed Forces of the United States as he determines to be necessary and appropriate in order to defend the national security of the United States against the continuing threat posed by Iraq; and enforce all relevant United Nations Security Council resolutions regarding Iraq." Cheney is more combative in his approach to these detractors, relishing the opportunity to point out statements of support the Iraq War resolution received from people like John Kerry, leaders who would later recant their support when it was politically expedient to do so.

In his book *Presidential War Powers*, Louis Fisher, a senior specialist in separation of powers at the Library of Congress, provides a history of the tensions within the nation's war powers. As he tells it, since the nation's inception there has been an ongoing battle between the executive branch and the legislative branch about who can declare war on behalf of the nation.

Updated in 2004 to provide a full treatment of the decisions to invade Iraq and Afghanistan, *Presidential War Powers* concludes, in essence, that the war of choice narrative is not helpful for examining the Iraq and Afghanistan wars since both sides of the debate share in the blame.

Fisher begins his argument by concisely stating a fundamental aspect of his premise: "the power to go to war was not left to solitary action by a single executive, but to collective decision making through parliamentary deliberations." Additionally, "the framers empowered the President to be Commander in Chief, but that title must be understood in the context of military responsibilities that Congress authorizes," he writes, citing seven clauses of the Constitution that place war powers with Congress. Even so, Fisher's history shows how these powers have slowly but steadily marched toward the executive branch with each use of military force starting in the 1800s.

Until World War II, Congress provided authorization for numerous uses of military force, and it wasn't until 1950 that the executive branch first ordered the military into action in Korea without congressional approval. Since 1950, the executive branch consistently won the struggle for authority, and it was further empowered by policy changes like the War Powers Resolution in 1973. By the time of the decisions to invade Afghanistan and Iraq, the struggle had evolved from being about legislative or executive supremacy to being about political positioning to take credit for military successes and to create distance from military failures by both branches.

Fisher challenges the war of choice narrative through his detailed history of how the war authorization discussion played out after September 11. Following the attacks, the president asked for and received authorization to use force against terrorists, but because the political climate had calmed and the case for invading Iraq was less clear, the president had to work harder to sell his decision to topple the Hussein regime. "Initially, the administration announced that President Bush did not need authority from Congress," Fisher writes. When that failed, they argued that "by passing the Iraq Liberation Act of 1998, they had already approved U.S. military action against Iraq for violations of Security Council resolutions," but this latter argument failed to inspire because the 1998 act was not binding and it did not explicitly authorize war, only money to Iraqi opposition groups. By October 2002, Congress passed the Iraq Resolution, which granted the president the authority to use force in Iraq; however, Fisher notes that Congress had not decided on war through this vote, rather, it decided only "that President Bush should decide."

Fisher is decidedly disgusted with Congress' actions during this time, calling them "distinctly junior varsity" and citing its task as no longer serving as an authorizing body, but rather simply endorsing "what the President had already decided." In passing the Iraq Resolution, he writes, "Congress failed to discharge its constitutional duties. . . . Instead of making a decision about whether to go to war and spend billions for a multiyear commitment, it transferred those legislative judgments to the President." By folding under political pressure, members of Congress "washed their hands of the key decisions to go to war and for how long" and "thereby abdicated [their] constitutional duties to the President."

Looking back at the long history of tensions between the coequal branches, Fisher writes that the Congress that authorized the war in Iraq "agreed to compromise language that left the decisive judgment with the President," which "was precisely what the framers hoped to avoid when they drafted the Constitution." The Iraq Resolution was "virtually identical" to the Gulf of Tonkin Resolution that left Rumsfeld uncomfortable, in that it transferred "to the President the sole decision to go to war and determine its scope and duration." In each case, lawmakers chose to trust in the President . . . they gave the president unchecked power." They did so by passing legislation that "would decide neither for nor against war."

What Fisher does not really answer in his study of war powers is how a sort of vote that would send a nation to war could result in such a political outcome. As a policy history, his book does not explore how it is possible for elected officials from two branches of government to make choices to deploy American forces while positioning to avoid responsibility for those decisions. One answer to this question can be found in a single policy change that almost entirely removed the American people from the extremely important foreign policy debate: the end of the draft.

During their political careers, and especially during the wars in Iraq and Afghanistan, many opponents had branded President Bush and Vice President Cheney as "chicken hawks," and both write about the subjects briefly and in very personal terms. About his college years, Bush writes, "the war in Vietnam was escalating, and President Johnson had instituted a draft," leaving men his age with two options, "join the military or find a way to escape the draft." His decision, in his words, was to serve, joining the Texas Air National Guard. Even though his service in the National Guard would be belittled during his presidential campaign, Bush concludes, "I would have been ashamed to avoid duty." Cheney never served in the military, choosing instead to pursue a graduate degree and securing numerous draft deferments.

Like Bush, he writes about his position on the subject clearly. If he had been drafted, he would "have been happy to serve."

Secretary Rumsfeld explores the subject considerably more. He had served in the navy before serving in Congress, but his opinion of the draft, like his views on presidential war powers, evolved. He writes, "the situation in Vietnam, and the demonstrations against the war and the draft, strengthened greatly my support for a transition to an all-volunteer military."

For Rumsfeld, the reasons for ending the draft were many. There were regular protests by young Americans who were "asking why they were being forced to fight in a war they did not understand and that they did not see as critical to our country's security." Draft exemptions seemed to benefit people with connections, making it so that "the draft also exacerbated racial and social tensions in the country." In 1967, Rumsfeld writes, he observed "one of the largest antiwar demonstrations in the Washington area," which held that "conscription was unwarranted, discriminatory, and unfair," and he found himself agreeing with the protestors.

In different ways, each realized the significance of the draft in the wars they oversaw, especially the war in Iraq. Attempting to counter criticism that he had not done enough to mobilize the nation for the war, President Bush writes, "short of a military draft—a step I strongly opposed—I'm not sure what more I could have done to encourage sacrifice." Cheney writes, "this was the first time since we'd created the all-volunteer force that our soldiers had been committed in an unpopular war. . . . As criticism mounted, we had to be absolutely clear, internally and publicly, that we would not compromise our fundamental mission for political reasons." And Rumsfeld notes that all of the Americans putting their lives on the line in their nation's defense were volunteering to do so, making their service nobler.

This was a unique political experiment in the nation, and one that would have extraordinary implications on the battle over war powers. The all-volunteer military outsourced the nation's fighting to a select few, it distanced the American public from the impact of the war, and it provided space for decision makers in both the legislative and executive branches to position away from accountability because such wars affected so few of their constituents. The president, his cabinet, Congress, international partners, and even Saddam Hussein and those who attacked the nation on September 11 could be blamed for the nation's choice to go to war, but only a small percent of the nation, those who volunteered to join the nation's military, actually decided to participate in that shared decision.

3

Cowboy George's Painless War

The Cowboy Narrative

IN A continuation of the war of choice narrative, the second Bush presidency is often accused of embracing a cowboy foreign policy after the September 11 attacks. The cowboy narrative suggests that Bush's worldview consisted of good versus evil, he was willing to go it alone, and he was willing to commit the nation to global, total war no matter the cost, and that this was exactly how he carried out his efforts. As with most narratives, though, the cowboy narrative is exceedingly appealing in emotional terms, but it is unfulfilling when considering Bush's political posturing contrasted with his actual execution.

The cowboy narrative is reinforced by Bush's pronounced beliefs, but it is complicated by his actions. After September 11, Bush would commonly announce his black-and-white worldview that pitted the nation against its enemies; throughout his presidency, he dismissed those who compared his wars to the Vietnam War, which he believed to be micromanaged and escalated due to President Johnson's ego; and he took great pride in positioning himself as the opposite of his predecessor, President Bill Clinton, who he and his advisors saw as Johnson's pathetic opposite—weak, poll-driven, and inviting attack. Bush embraced the gray area within his black-and-white worldview, allowing him to execute in a style remarkably similar to the derided positions of Bill Clinton while achieving Johnson's devastating results. This can be most clearly seen through his tortured decisions to put and keep ground forces in both Afghanistan and Iraq while remaining unwilling to take casualties or ask for other significant sacrifices from the nation.

The one author that covered the tension between this narrative and reality more than any other during Bush's eight years in office is arguably Bob Woodward. Woodward's fours books on the Bush administration at war provide a framework for understanding the complexity of the positions taken by the Bush administration that drove the nation to war; context for how those decisions were actually carried out in practice; and, often through the journalist's direct questions to the president, they offer details that complicate claims that Bush was a cowboy bent on prosecuting a total war.

Throughout Woodward's four books on the Bush administration at war, the distinction between rhetoric and action is clear. The gulf between rhetoric and action provided a gray area that allowed the president to talk tough publicly but to continually make decisions that looked more like limited war. The story of a brash foreign policy juxtaposed with a coddling domestic policy are, subtly, the centerpieces of Woodward's four books, *Bush at War*, *Plan of Attack*, *State of Denial*, and *The War Within*, and Woodward's conclusions are reinforced by other authors who are critical of the Bush administration's efforts.

As Woodward points out in *Bush at War* about the president's immediate response to the September 11 attacks, the president clearly embraced the cowboy narrative. Woodward describes a time shortly after the attack when the president, addressing a group of military reservists, states the "bad guys" who attacked America were "Wanted Dead or Alive," like on "an old poster out West." Woodward also describes a private conversation Bush had with King Abdullah of Jordan where Bush was tempered with what Woodward calls Bush's "dual impulses." "We're steady, clear-eyed and patient," Woodward quotes Bush as saying to the king, "but pretty soon we'll have to start displaying scalps."

These dual impulses meant that balancing the response to the attacks with the national desire for vengeance was complicated. "My attitude all along was, if we have to go it alone, we'll go it alone; but I'd rather not," Woodward quotes the president. But at the same time, there were limits to what he would ask of the nation. "A month from now, Americans will be watching football and the World Series," he says, "but the government would have to carry on the wars indefinitely."

The closer to the September 11 attacks they were, the more cavalier the Bush administration was about the national costs of war. Woodward tells of CIA official Cofer Black, who noticed immediately the administration was "moving from the bureaucratic mode to acceptance of risk, even death." Later, Woodward tells of the cost being openly discussed, with Black explaining to the president, "you've got to understand, people are going to die." Bush's response was simple: "that's war."

In the immediate internal deliberations around the necessity of ground forces in Afghanistan, Woodward describes the evolution of this discussion as beginning broadly, with Defense Secretary Rumsfeld stating, "maybe we should put some people on the ground." Such forces could be used "to conduct humanitarian operations both in the north and the south," the secretary suggested, allowing them to show that they did not intend to be occupiers, and "help mute some criticism of our operations." Additionally, putting boots on the ground would "start creating an environment in which Afghanistan becomes inhospitable to the al Qaeda and the Taliban." For his part, Bush agreed. He did not want the response to appear to be one of conquest and occupation, either. "I want a humanitarian aid drop in the north and the south," the president said, asking, "can we have the first bombs we drop be food?"

These impulses were further complicated by the administration's desire to not respond to the attacks through what they called the "Clinton option." As they saw it, Clinton responded to national security threats by ordering safe air strikes to simply "pound sand," which was a posture "so weak as to be provocative, a virtual invitation to hit the United States again." In response to the September 11 attacks, the Bush administration was determined not fight a "photo-op war," as Bush stated, but rather to give the American people the "big bang" they wanted. This sort of response served a diplomatic purpose as well. "Tell the Afghans to round up al Qaeda," Woodward cites the president as saying. "We're going to hurt them bad so that everyone in the world sees, don't deal with bin Laden," Bush said. "I don't want to put a million-dollar missile on a five-dollar tent."

The internal discussion led to the development of military options for Afghanistan. According to Woodward, the first review of contingency plans left plenty to be desired, according to General Hugh Shelton, who had also been President Clinton's chairman of the Joint Chiefs of Staff. Ordered to update these plans, Shelton presented three new options, two of which sounded like the "Clinton option," but the third included minimal air and ground forces.

The development of these options mirrored new rhetoric from the president, who began publicly stating that he was determined to use all elements of national power to defeat terrorism. The president said, "everything is on the table," and as the nation's commander and chief, he saw it as his responsibility "to do whatever it took to win," even putting the lives of Americans at risk in order to "rain holy hell on" the enemy. Addressing a television audience of eighty million viewers on the response to the attacks, Bush offered a clear refutation of Clinton's foreign policy, declaring that his administration would "direct every resource at our command," including "every means of diplomacy, every tool of intelligence, every instrument of law enforcement, every financial influence, and every necessary weapon of war, to the disruption and to the defeat of the global terror network." This powerful language provided the nation what it wanted to hear, but it was not how the war in Afghanistan would be waged.

CIA operatives were in Afghanistan almost immediately after the attacks, but military forces would take more time, initially because military leaders refused to deploy them until they had established Combat Search and Rescue, or CSAR, assets, according to Woodward. The president acknowledged the need for CSAR, stating, "it's very important to realize how do you balance the military's desire to cover all contingencies at least once, maybe sometimes twice—they're relatively risk-adverse and they should be, after all they're dealing with people's lives—versus the need to, for whatever reason, to show action." Even so, he continued challenging his leaders to do more in less time, reminding them that he was not interested in launching "Clinton like cruise missiles into tents."

Unfortunately for the president, even as the CSAR capability developed, another problem became clear. Afghanistan only had "mud-hut-type targets," which did not justify the risk of deploying troops. Rumsfeld eventually presented a solution that would meet everyone's needs: they would bomb from northern Afghanistan, but in order to do so safely and without CSAR, they would be using B-2 bombers and cruise missiles. As Woodward puts it, "the plan had just ever so much a ring of how Clinton might do it—safe, less than optimal, a compromise."

When the air campaign started the targeting problem persisted. Within three days, all of the planned targets had been destroyed, leading the media to press for details on how the military moved forward without any targets. Rumsfeld responded with a joke, stating, "we're not running out of targets, Afghanistan is." Woodward describes the president as seemingly "exhibiting some patience" for the moment. He was "glad finally to be bombing." But when he later asked the president about this, the president responded that they only seemed to be pursuing his predecessor's approach: "We're just bombing sand. We're pounding sand," he said.

To move beyond the "Clinton option," they needed ground forces, but initial plans showed they would need around fifty thousand. Since that number was too high to gain traction, Rumsfeld pushed for better plans. Finally, Rumsfeld was able to put special operators in-country and able to announce "the campaign had moved beyond bombing, that U.S. forces were on the ground."

As the initial challenge with CSAR points out, one significant reason for the delay of ground forces was a "zero casualty" mentality. While the president spoke of the "bad guys" being "wanted dead or alive," he was also aware of how quickly the national mood could be swayed by what General Shelton called the "Dover test." This notional test measured the nation's limited tolerance for the flag-draped coffins returning to the Dover Port Mortuary during wartime.

The "Dover test" exemplified the true challenge of the president's dual impulses. Bush's narrow respect for the nation's resolve limited his willingness to deploy ground troops to Afghanistan, which led to the growing perception that the United States had been quickly stalemated. There were regular comparisons to the Vietnam War, even with talk of "Americanization" of the war before it became a "quagmire." Conservative commentators Bill Kristol and Charles Krauthammer were even decrying the softness of Bush's approach as being "flawed" and "fought with 'half-measures,'" respectively. Fortunately for the Bush administration, the situation in Afghanistan improved in November 2002, quelling the criticism momentarily.

The momentary victory in Afghanistan did more than silence detractors, though, it also allowed for an evolution in the administration's perspective on the use of military force. "In all," Woodward writes, "the U.S. commitment to overthrow the Taliban had been about 110 CIA officers and 316 Special Forces personnel, plus massive airpower." They could deny that they were pursuing the "Clinton option," but in terms of actual risk and commitment, their approach was remarkably similar.

The president had talked total war but pursued limited war in almost every way. Woodward asserts that the president "had not put the country on a war footing, demanding sacrifices from large numbers of citizens, or taken what for him would be the unthinkable and draconian step of raising taxes or repealing his 2001 tax cut." Woodward asked the president, "was it not possible that he had undermobilized given the threat and the devastation of September 11?" The president's wandering response covered increased spending for military, intelligence, law enforcement, and first responders, and little else. The early success in Afghanistan confirmed that the nation could win wars cheaply, which informed the thinking around the invasion of Iraq, a war that would only magnify the president's impulse to rhetorically embrace the cowboy narrative while not engaging all elements of national power to win.

Plan of Attack, Woodward's second book on the Bush administration at war, discusses the lead-up to the war in Iraq following initial success in Afghanistan. Woodward writes that early on, Paul Wolfowitz, Rumsfeld's deputy, was the only advocate for war in Iraq. But after the intoxicating success in Afghanistan, as Woodward describes it, Bush and Rumsfeld began to evolve. Bush developed the Bush doctrine, which expressed his black-and-white foreign policy while he quietly executed in the gray area. For Rumsfeld, Afghanistan confirmed his desire to transform the military into a leaner, quicker fighting force by leveraging technology and preventing casualties. Rumsfeld's vision, in other words, offered an operational complement to Bush's dual impulses.

According to Woodward, Rumsfeld saw the invasion as an opportunity to "tune up the war machine everywhere," an assertion that is supported in the 2006 book *Cobra II* by Michael R. Gordon and General Bernard E. Trainor. This book, an extraordinarily comprehensive history of the lead-up and initial execution of the war in Iraq, focuses on the administration's early policy decisions and how they contributed to much of what followed during the invasion and the early days of the occupation. An inescapable conclusion of *Cobra II* is that the insistence on going to war in Iraq, and doing so with as few troops as possible, led to a chaotic invasion and an even more challenging reconstruction period to follow.

According to *Cobra II*, Rumsfeld ordered General Tommy Franks, the commander of Central Command, to update the existing plans for Iraq, which dated back to the late 1990s and were developed by General Anthony Zinni, the previous commander of Central Command. To Rumsfeld, Zinni's Iraq war plan was an enormous piece of bureaucratic literature, and worse, it called for a massive deployment of force of over four hundred thousand troops, which would also require a lengthy deployment period before the invasion could begin. This approach was clearly not in line with Rumsfeld's vision of a transformed US military that could deploy quickly and win with a small number of forces.

Franks's first effort to rewrite Zinni's plan only trimmed one hundred thousand troops, which was equal to the minimum number of troops that the initial plan called for. Rumsfeld rejected the proposal, arguing for the plan to embrace his new paradigm instead of just refinements on the old plan. Franks's next effort to reduce the number of forces, and the time it would take for them to be in place, was more successful, reducing the required troop levels to three hundred thousand and the deployment timeline to 120 days. The updated plan relied on "lines of operation" that included "airpower, Special Operations Forces, conventional ground forces, information operations, and civil-military operations," each line of operations allowing Franks to influence different aspects of the military and political landscapes with fewer troops and in a shorter time period. While the new approach was again rejected by Rumsfeld, Franks had shown him a model for how deploying fewer troops in less time could be applied to reduce those numbers even more. That is, when assumptions were weighted differently, the lines of operations allowed them to manipulate categories to manage the troop strength and the deployment timeline. Under the new plan, they could invade Iraq with 145,000 troops and grow that number to about 275,000 within months.

Even though Franks had halved the original numbers and provided a plan for a rolling start to the invasion in order to expedite troop deployment, Rumsfeld challenged Franks to use lines of operation to continue to reduce the new numbers, and Franks complied. His planners would look at the possibility of relying on deployments from partner countries, further tweaking the lines of operation, and any other measures that would allow them to fit the realities of a massive invasion into Rumsfeld's vision for transformation. Eventually, Franks was able to brief a plan that relied on even smaller force numbers and a faster deployment period that would be capable of overthrowing the regime in 135 days. Even though Franks was briefing this plan mere months before the invasion began, Rumsfeld pressed him to reduce the timeline and troop numbers even more.

While Rumsfeld was setting his war machine into motion, the president was settling on a strategy of coercive diplomacy for dealing with Saddam Hussein by offering diplomatic engagement as an option, but the military would address noncompliance. As Woodward points out, there were problems with this dual approach. Bush "was planning for war, and he was conducting diplomacy aiming to avoid war." As was the nature with coercive diplomacy, there were times when "the war planning aided the diplomacy; at many other points it contradicted it."

As with coercive diplomacy, there were also contradictions between Bush's actions and his rhetoric, contradictions that he used to his advantage. During his State of the Union address in January 2002, Bush coined the phrase the "axis of evil" to describe Iraq, North Korea, and Iran. Although the vague phrase caused confusion, one thing was clear according to Woodward: "The go-get-'em cowboy talk now had put three countries in his sights, most provocatively his father's nemesis, Saddam Hussein." More important than this clear declaration, though, was how Bush intended to use the phrase to his advantage. "It served a dual purpose for Bush," Woodward writes. First, "it sounded tough," but it also "blurred the focus by including North Korea and Iran, providing additional cover for the secret planning for covert action in Iraq, and war."

Bush would continue this approach in future speeches, which were primarily written by Michael Gerson. Woodward points out that "the perennial debate in foreign policy between the realism of Teddy Roosevelt's 'big stick' and Woodrow Wilson's idealistic goal of 'making the world safe for democracy' was sterile in Gerson's view." According to Woodward's analysis of the speechwriter, "a president needed realism *and* idealism, and Gerson believed Bush wanted both and to be able to say, in effect, we take power seriously, we take ideals seriously." This approach manifested when Bush delivered a speech on Iraq to the United Nations. "The soft-liners liked it because the president was seeking international and United Nations support," Woodward writes. "Hardliners liked it because it sounded tough."

The contradictions inherent in coercive diplomacy and between Bush's rhetoric and his willingness to commit the nation to war had significantly affected Franks's planning efforts. Realizing that the previous plans were flawed, they opted for a fresh start, this time developing an option that allowed for "a shorter time between the president's decision to attack and the attack itself." This would allow them to deploy fewer troops more quickly, which aligned with Rumsfeld's vision, and without much detection, which aligned with the president's position that he was still pursuing diplomatic efforts. According to Woodward, Franks's new plan could have fifty thousand troops in Kuwait to begin the invasion within three weeks and another fifty thousand on the ground in the following weeks. This aligned with Bush's plan and rhetoric as well as Rumsfeld's focus on transformation.

As the Iraq decision approached, Woodward shows that the president's advisors were in agreement. Rice, who was a proponent of coercive diplomacy, advised that the president had to follow through. Rumsfeld warned that relationships were at risk if the president did not invade. But Cheney leveled the toughest argument, stating, "once the president had laid out his objective of regime change, and begun the process of troop deployments and CIA work, then if they didn't follow through, they would be like Clinton—a lot of bold talk and not much action." But in fact, the president had already made his decision to invade, which left him "enveloped in a contradiction," Woodward writes. "He had privately decided on war, but publicly he was continuing diplomacy." This was not new, it was "a world of more dilemmas, dissonance, and hedging."

With the decision made, Rumsfeld deployed troops in small packages of ten thousand to twenty thousand. "By the middle of February," Woodward writes, "there would be a total U.S.-force level of 140,000 in the region, 78,000 of which would be ground forces—Army, Marines and Special Operations Forces." Before the invasion, forces in the region had grown to 241,516, with over 40,000 more coming from coalition partners, and on March 21, marines invaded Iraq. "Resistance was light," Woodward writes. "The first casualties were four American and eight British troops who were killed in a helicopter crash."

The first casualties of the war were important, not only because of their sacrifices but also because of the challenge that war casualties would present. Part of the "Clinton option" was the belief that Clinton had been unwilling to risk casualties, but the Bush administration proved that it was just as sensitive. The similarity between the two administrations was not only a contradiction of their dismissal of Clinton, but it also directly challenged Secretary Rumsfeld's vision for the military and the president's rhetoric of resolve.

Throughout the planning process, low casualties seem to have been embraced as another outcome of America's military might, which had been proven in Afghanistan. Woodward cites a discussion where the prospect of a longer war with increased casualties was used to push up the invasion timeline. Another passage discusses the "fortress Baghdad" scenario, or the prospect of Saddam leaving an insurgency in his wake, which "would be a military calamity that could mean a long war with high casualties." Franks even privately predicted "there would be fewer than 1,000 casualties on the U.S. side and probably only several hundred."

It was not long before they and the rest of the nation started to realize that Franks's calculation was incorrect. Woodward describes the period following the whirlwind invasion as "a dark period for Bush and the team with the Army stalled, dozens of Americans killed, some taken prisoner and the news coverage negative." Violence was increasing and it became clear that the "fortress Baghdad" was now a reality.

Within a year of the invasion, Bush's key advisor saw the political downside of the war, especially during an election year. "The U.S. military had more than 100,000 troops there and would require that many more for some time," Woodward writes. "American soldiers were being killed at too high a rate," he adds, "and they hadn't reached a political settlement."

The early days of the war in Iraq presented an application of the "Dover test," resulting in a crippling situation for the Bush administration. If the nation continued to take casualties, it would lose interest, but the only way to stabilize Iraq was to introduce more ground forces. While crippling, this situation was also the direct result of Bush's public tough talk that rallied the nation contrasted with his lack of commitment to actually deliver on his promises. Woodward describes these challenges in his next works on the Bush administration's wars, *State of Denial* and *The War Within.*

The calm following the invasion was broken on Memorial Day, 2003, with the death of the army private Jeremiah D. Smith from Missouri, and the attacks would only continue to tick upward from there. "Classified reports showed that the insurgent attacks had jumped to 1,000 in the month of October, more than 30 a day," Woodward writes. By March 2004, the situation reached sickening new heights when four Blackwater contractors were killed, burned, and left hanging from a bridge in Fallujah.

Numerous books reflect on the consequences of Bush's unwillingness to send enough troops into Iraq and Rumsfeld's insistence on fighting with a lean, transformed military. This issue was an early part of the public debate after the situation in Iraq deteriorated immediately after the invasion, and it continued on after, becoming a sustained criticism of Bush's handling of the war. Thomas Ricks's 2006 book, *Fiasco,* names troop strength as a central weakness of what he calls "perhaps the worst war plan in American history," while retired Lieutenant General Daniel Bolger cites the administration's focus on low troop numbers as a key to America's defeat in Iraq and Afghanistan in his 2014 book, *Why We Lost.*

In *Fiasco,* Ricks opens by outlining the common aspects that contributed to the Bush administration's assumptions and policies about Iraq. He explores the history of American involvement in Iraq and the growing belief by neoconservatives that Clinton's decision to merely contain Hussein as disastrous. One critic of the Clinton administration, Danielle Pletka, even goes so far as to assert that Clinton and his team were exceptionally "casualty averse," and that their approach had weakened American foreign policy. While Bush had vowed military restraint while campaigning for the presidency, the September 11 attacks created an opening for perspectives like Pletka's to dominate.

Ricks describes an inadequate invasion force that was even more ill-suited to serve as an occupation force. He argues that the private disagreements about troop levels broke into the public discussion about the war following the first major disaster of the invasion: an ambush of thirty-three American soldiers who had been cut off from convoy, resulting in eleven dead, nine wounded, and seven captured, including Jessica Lynch. The ambush was only four days into the ground war, and the invasion would only get more complicated going forward.

This setback, and the others that followed, immediately brought criticism from retired generals and public intellectuals about the inadequacy of the troop levels in Iraq. While Pentagon officials rebuffed these criticisms as being unpatriotic and ill informed, Rumsfeld and his cadre of military transformers doubled down. Not only was the invasion force adequate, but the additional forces called for in the approved plan would be reduced as well, arguing that it would take fewer forces to occupy than to invade.

Bolger's *Why We Lost* serves as a remarkable reflection from a senior military leader following the wars in Iraq and Afghanistan. In it, he accepts the blame for the defeats for himself and for his fellow military leaders. He and his fellow generals lost the wars, he states, because they failed to stand up and advocate for the forces they led. They lost because they allowed centuries of military learning, experience, scholarship, and strategic norms to be overwhelmed by ideas like Bush's focus on nation building and Rumsfeld's transformation.

The failure to advocate for basic military principles covered nearly every aspect of the war, but especially with respect to troop levels. As the Iraq War became more complex, the lack of troops forced an acceptance of politically derived objectives and policies. In order to win in Iraq without additional troops, the military would rely more heavily on National Guard and Reserve component forces; the administration would make additional pleas for troop deployments from allied nations; and, the focus for leadership on the ground in Iraq was driven toward rebuilding and empowering the Iraqi military to take the lead. Only after these efforts failed, and when the political landscape in the United States was just as inhospitable for the Bush administration, did some seize on the idea to deploy more troops, but by that time it was already too late.

But long before the surge was contemplated, Bush had to win reelection. In an election year, there was no political benefit in running away from the war that defined his presidency, so Bush continued his familiar rhetoric. "There are some who feel like that if they attack us that we may decide to leave prematurely," Woodward quotes the president saying to a reporter. And after a testy exchange with the reporter, Bush added a bit of his cowboy flare. "My answer is 'Bring 'em on.' We've got the force necessary to deal with the security situation."

Although this statement was "ill-advised," as Woodward describes it, pronouncements on his foreign policy would only intensify as the election approached and the conditions on the ground worsened. Woodward quotes Bush cheerleading at internal National Security Council meetings carrying the same message as well: the difficulties on the ground should strengthen their "resolve to do what we have to do for freedom," and "groups that respond by pulling out of Iraq are simply giving in to the killers and rewarding them," he said. In Bush's words, America would "stay the course in Iraq."

At other times, Bush argued that the attacks were actually a good sign, stating the "progress makes the remaining terrorists even more desperate and willing to lash out against symbols of order and hope, like coalition forces and U.N. personnel." He would speak publicly of "not only containing the terrorist threat" but also "turning it back," even while the number of enemy-initiated attacks was climbing. When his argument for staying the course grew to include honoring the fallen, some within his own administration were outraged. As Secretary Powell's lieutenant Richard Armitage stated to the secretary, "another way to say that is we've got to have more men fall to honor the memories of those who have already fallen."

The standoff in Fallujah in April 2004 following the death of the Blackwater contractors provided a painful glimpse of Bush's predicament: how to choose between casualties and troop levels. Insurgents had taken over the holy city, and American forces had surrounded it, only to be ordered to abandon under pressure from the Iraqi Governing Council. Ambassador Paul Bremer visited Fallujah during the standoff, noting that the stalemate was "the most dramatic example" that there were not enough troops. Kurdish President Barzani was furious that the American forces stood down. Explaining the decision in Clintonian terms, Barzani stated that after Fallujah, every Sunni insurgent knew that American forces would back down if they were faced with enduring casualties, Woodward writes.

Administration officials tried to figure out new ways to meet the mission in Iraq without taking additional causalities and without additional forces. Secretary Powell wanted to recruit a new military from the mid-level commanders of the Iraqi Army that Bremer had disastrously disbanded, but Rumsfeld disapproved. General John Abizaid, the new commander of Central Command, recommended they work to embed American troops with Iraqi units to support the fledgling forces and to develop opportunities for intelligence. The president loved the idea, but Rumsfeld again disapproved, stating the possibility that the embedded American troops might be killed by their Iraqi comrades.

Fortunately for Bush, the disparity between his actions and rhetoric did not hurt his chances for reelection, thanks to what Woodward cites as the fact that Senator John Kerry appeared "uncertain and indecisive, while Bush succeeded in presenting himself in the campaign as consistent and tough." Following the election, Woodward writes that Bush met with Gerson to discuss his new vision. "Bush's goal now was to dramatically alter the American foreign policy mind-set as radically as it had been changed at the beginning of the Cold War in the late 1940s with policies of containment and deterrence." Kennedy said that America would "pay any price, bear any burden, meet any hardship, support any friend, oppose any foe to assure the survival and the success of liberty," but to Gerson, "that was the mindset that led to Vietnam," and he "did not want something that suggested some wild-eyed, open-ended commitment to full democracy everywhere." During his second inaugural address, Bush stated, "it is the policy of the United States to seek and support the growth of democratic movements and institutions in every nation and culture, with the ultimate goal of ending tyranny in our world."

While the president articulated his new vision, the situation in Iraq deteriorated further. After having been reelected, his team suggested he reshape his message to make room for possible changes in policy. Bush did not like the changes from his long-standing message of American resolve, but he eventually accepted them. But even so, "it would take time to wean the president from his zero-defect proclamations."

National patience was wearing thin, and Bush's political opponents saw an opportunity to attack, with Pennsylvania congressman Jack Murtha introducing legislation to bring American forces home. The Republican response was straight from the Bush playbook, with Speaker Dennis Hastert accusing Murtha and others of wanting to "cut and run," and the president's press secretary stated that Murtha's bill amounted to Democrats wanting to "surrender to terrorists." The bill died.

It was not just American casualties that drove the politics, but also Iraqi casualties. One reporter pointed out that Iraqis "suffered the equivalent of a 9/11 attack every week," while the Iraqi interim prime minister, Ayad Allawi, noted that between fifty and sixty Iraqis were killed each day. To one question about the number of Iraqis killed, Bush responded, "I would say 30,000 more or less, have died as a result of the initial incursion and the ongoing violence against Iraqis." But more telling was when he followed this estimate by saying, "we've lost about 2,140 or our own troops in Iraq." Woodward notes that the president "had almost the precise death toll for Americans, 2,144, right on the tip of his tongue."

"President Bush loved to keep scorecards," Woodward writes. The National Security Advisor "prepared a situation report each morning for the president and invariably the first item was Iraq and the [American] casualties." There were two more situation reports provided later in each day. His military advisors detested body counts because of how they were used in Vietnam, but the reports kept coming in because "meeting the president's emotional and political needs was apparently more important."

While the situation on the ground was getting worse, some thought bold measures were in order. National Security Advisor Stephen Hadley and his director for Iraq Meghan O'Sullivan set out to develop a public relations effort to change the administration's casualty-averse approach to war, and they brought on Peter Feaver to assist in June 2005. Feaver, an academic with an expertise in "the impact of war on public opinion," had been a staffer for Clinton's National Security Council, where he believed that Clinton had created a culture of "virtually no tolerance for casualties." Feaver's research showed that the American people were less casualty averse than their leaders, and his work informed Hadley's and O'Sullivan's approaches.

Meanwhile, continued casualties affected key officials differently. Rumsfeld thought the rising casualty numbers provided negotiating power with the Iraqi government. Woodward cites a Rumsfeld memo titled "U.S. Casualties" that outlined his case. "You could tell them that the longer it takes them to get a government and the longer it takes them to start providing leadership," Rumsfeld writes, "the more people are going to be killed," adding, "there has to be a limit." Bush responded by monitoring his body count scorecard, which at least allowed him "to know whether or not our soldiers are fighting."

In addition to the efforts of people like Hadley, O'Sullivan, and Feaver, the Iraq Study Group began meeting to discuss whether the United States should send in more troops or bring them all home. When it questioned administration officials on the need for more troops, the answer was consistently negative. Woodward reports that member Chuck Robb realized that they were "coalescing around a recommendation for a fixed early withdrawal from Iraq," when he penned a memo against that option. Instead he called for "a significant short-term surge in U.S. forces on the ground." He wrote, it was "time to let our military do what they're trained to do on offense—without being overly constrained by a zero casualties or collateral damage approach." As Robert Gates would later relate in his memoir, *Duty*, Robb's recommendations were buried deep within the Iraq Study Group's report.

The argument was compelling, but they would still need to convince the president. In his interview with the Iraq Study Group, former President Clinton suggested they appeal to his political self-interest, stating, "if the president announces a change in course, he will get more time—if you can convince him to change," Woodward quotes Clinton. The president eventually agreed, removing Rumsfeld immediately after the election, replacing General Casey with General David Petraeus, and ordering a surge of temporary forces into Iraq with the hope of stability for long enough that the Iraqi government could realize a political solution.

It was a monumental change in the president's strategy, and it untied the hands of his military leaders for the first time. Previously, he had been committed to a zero-casualty solution, but with his decision to surge, he was not just renouncing this approach, but doing so in a way that guaranteed increased casualties. He made this decision even while his top general argued surging troops would only have a "temporary, local effect."

The service chiefs realized immediately that filling the need for five additional surge brigades would be the military's problem alone. "We have forgotten how to use the other elements of national power except for the military," Woodward quotes the air force general Michael Moseley. General Peter Schoomaker, the chief of staff of the army, noted, "the nation needs to mobilize," adding "most of the country is in spectator mode." Marine Corps commandant General Michael Hagee said they "should have asked the American people to sacrifice by imposing a 5-cent gas tax on every gallon of gas or something like that after 9/11." Taxes, draft, and any other support were absolutely out of the question, they realized.

In *Hubris: The Inside Story of Spin, Scandal, and the Selling of the Iraq War* by Michael Isikoff and David Corn, the president is quoted as being reflective about his cowboy rhetoric. When asked to identify a mistake he had made with respect to the Iraq War, Bush responded in general terms about tactics that could have been better, and then he identified his own messaging. In part, Bush said, "Saying 'bring it on,' kind of tough talk, you know, that sent the wrong signal to people." Bush added, "I learned some lessons about expressing myself maybe in a little more sophisticated manner—you know, 'wanted dead or alive,' that kind of talk." But these generals' concerns suggest that they were far less worried about his rhetoric than his unwillingness to assume political risk to mobilize the nation in a meaningful way, even if doing so meant accepting that achieving the goals of his invasion of Iraq could not be accomplished quickly or without sustaining casualties. That was war, as Bush had said early on, and there was no way to have war without the nation and its military enduring pain and sacrifice together.

4

A Phase IV Rodeo

The Phase IV Narrative

A PERSISTENT Iraq War narrative asserts that the Bush administration lacked a Phase IV plan—the plan would outline reconstruction efforts in Iraq following the invasion in Phase III. Following the toppling of the Hussein regime in April 2003, the Phase IV narrative was seemingly confirmed by events on the ground: looting, discord, and the rise of the insurgency. By late 2003, it seemed clear that the war plans ended with a conquered Baghdad.

The Phase IV narrative is an outgrowth of previous war narratives. If the Bush administration engaged in a war of choice, then they acted with arrogance. Their arrogance led them to believe in the possibility of a zero-casualty war. And after this painless invasion, the Iraqi people would welcome the Americans as liberators, making planning for occupation and reconstruction unnecessary. But the Phase IV narrative places too much emphasis for Phase IV planning on senior administration officials, it overlooks the planning processes that are habitually carried out by military leaders and other officials, and, as with most narratives, it mostly serves as a political attack.

In fact, there were plans for the reconstruction phase, but what was lacking was leadership and coordination. The Phase IV planning efforts lacked a clear vision for victory and how to achieve it. They lacked a commander in chief who clearly stated whether the invasion would lead to redeployment or nation building. And they lacked an administration that was flexible enough to evolve with the threat it faced. Even though there were numerous plans for Phase IV, the lack of leadership and coordination made it so the important reconstruction effort was carried out aimlessly.

An interesting outcome of the war in Iraq is the small number of books written from the perspective of those who participated in the Iraq reconstruction effort, and the books that do exist only begin to cover the reconstruction effort. *Reconstructing Iraq*, by Gordon Rudd covers the story leading up to the establishment of the Coalition Provisional Authority, or CPA. Paul Bremer's *My Year in Iraq*, and James "Spike" Stephenson's *Losing the Golden Hour* tell competing versions of efforts during the CPA's time. Robert Earle's *Nights in the Pink Motel* tells of the planning that took place during the General George Casey and Ambassador John Negroponte era that followed the CPA. And *We Meant Well*, by Peter Van Buren, tells of the efforts to reconstruct Iraq during the so-called civilian surge near the end of the war. Each of these books covers only limited aspects of various periods of the overall reconstruction effort in Iraq, but *Hard Lessons*, the official report of the Special Inspector General for Iraq Reconstruction, or SIGIR, provides an invaluable backdrop for the overall reconstruction effort.

Hard Lessons covers the SIGIR's inspection of "about $50 billion in U.S. funds appropriated by the Congress for Iraq—the largest relief and reconstruction effort for one country in U.S. history." The report identifies five persistent failures that contributed to the demise of the reconstruction effort: how security challenges prevented reconstruction, frequent and reactive course corrections to the reconstruction strategy, continual personnel turnover, systemic waste, and poor unity of command. These themes and challenges only evolve throughout the SIGIR's coverage of the war, they never resolve.

Starting in the first chapter, "Planning Begins," the SIGIR provides countless examples of the leadership and coordination failures that led to the failed reconstruction effort, mainly involving the Departments of State and Defense and starting as early as the fall of 2001. According to the SIGIR, Defense viewed the war as a form of regime change where "power should rapidly transfer to an Iraqi authority," which led to a view of "U.S. forces as liberators who would leave Iraq within months of toppling the regime." Conversely, State "concluded that invading Iraq and replacing its totalitarian regime would require a U.S. commitment of enormous scope, carried out over a period of years, engaging everything from Iraq's judiciary to its electrical grid." The SIGIR concludes, "the intense interplay between these competing visions fundamentally shaped the process of prewar planning for postwar Iraq in the fifteen months that preceded the March 2003 invasion."

The State Department established a massive planning effort called the "Future of Iraq Project," which would "serve both as a means to expand postwar planning and as a vehicle to consolidate competing Iraqi exile groups, who were involved with the project from its earliest days." The White House's plans reflected the Defense Department's vision and began in the spring of 2002 through twice weekly National Security Council's deputies' meetings. During these meetings, members focused on three possible approaches: "a liberation model in which Iraqis would quickly take charge through a provisional government; a military administration led by Central Command; or a civilian transitional authority, perhaps run under UN auspices." By mid-2002, the SIGIR notes, none of the major difference had even been addressed, nor had a unified planning effort started between State and Defense.

In August 2002, approximately seven months before the invasion, President Bush signed a national security document titled *Iraq: Goals, Objectives, Strategy*, yet additional planning groups developed. By the end of that month, the SIGIR writes, "the Joint Staff instructed CENTCOM that it should begin planning to administer Iraq for an interim period after the invasion." National Security Advisor Condoleezza Rice established a new planning committee within the NSC called the "Executive Steering Committee on Iraq." And around the same time, Douglas Feith, Rumsfeld's deputy, established the "Office of Special Plans" within the Pentagon "to engage in large-scale planning for the war." Rumsfeld directed Feith to lead yet another postwar planning office in October 2002, but canceled the request a few days later. Because the president did not want to reveal his war plans while pursuing diplomatic solutions ahead of the invasion, the various planning efforts continued in secret, separately, and without coordination.

As the administration moved closer to war, the lack of real coordination resulted in planning failures. The deliberations for the Department of State's "Future of Iraq Project" went into 2003, preventing its inclusion in the Pentagon's plans, and it was over a thousand pages, leading the NSC steering committee to conclude that it "did not look like an operational plan." Meanwhile, the USAID, which is generally responsible for reconstruction efforts, developed its *Vision for Post-Conflict Iraq* in late 2002 without ever being consulted. The SIGIR writes, "the marked separation between civilian and military preparations, which had existed since late 2001, was followed by further fragmentation within interagency planning processes, which had begun in earnest in August 2002." The report concludes, "even as officials thought they were moving toward an integrated master plan, the building blocks of that plan were being developed in a piecemeal that rendered risks and needs less visible."

As these plans developed separately, the worldviews of the planners persisted. The effort loosely coordinated by the NSC was clearly becoming a military-dominated effort, meaning that the Defense view drove the process. As the SIGIR puts it, the Defense effort relied on a best-case scenario for reconstruction planning, meaning that the American investment in the reconstruction was limited because Iraqi liberation would be paid for through oil revenues. At the same time, the USAID-led effort was preparing for a humanitarian worst-case scenario. Again, it appeared that a leader was lacking who could see this "striking asymmetry," as the SIGIR puts it.

By late 2002, the Pentagon began increasing its planning apparatus, though not its coordination. A forty-person team led by an army major initiated an intensive planning session in late December 2002. Lieutenant General George Casey, then the director of the Joint Staff, created an additional planning cell of fifty-eight personnel called Joint Task Force 4, or JTF-4, to augment the Central Command planning effort. About a month later, and less than two months before the invasion, the president formally put Rumsfeld in charge of postwar Iraq and created yet another planning and managing organization, the Office of Reconstruction and Humanitarian Affairs, or ORHA, by signing National Security Presidential Directive 24.

In *Hard Lessons*, the SIGIR defines the ORHA period as one of mostly continued and focused dysfunction preceding the war. On January 9, 2003, retired Lieutenant General Jay Garner was asked to lead the new organization, moving into office fifty-six days before the war in Iraq began. As the SIGIR notes, Garner's involvement was based on the idea that he would activate the existing Phase IV plan, but "no master civilian plan was there to greet him." Beyond a missing plan, they also lacked basic accommodations, including email capabilities, and the staff he assembled was largely based on those he could convince to join the effort, all of which the SIGIR report describes as "a portrait of things to come in post-invasion Iraq."

Complicating matters further, ORHA remained a duplicate agency, not a coordinating one. The SIGIR notes that Garner did not know of Feith's Office of Special Plans until it was too late, that JTF-4 was pursuing a similar mission for Central Command, and no organization had clear authority. As the SIGIR puts it concisely, "in a period of three weeks, the Pentagon's military and civilian leadership had launched what became competing postwar organizations."

This competing structure, with ORHA and Central Command both under the Pentagon, mirrored the external rivalry between Defense and State. The SIGIR notes that a month before the invasion, military and civilian leaders involved in postwar administration met for a rehearsal, but there were no senior State officials present, and major military entities were given instructions that "limited the extent of their cooperation with ORHA." The rehearsal was also where Garner learned of State's Future of Iraq Project for the first time.

Gordon Rudd's *Reconstructing Iraq* describes the ORHA period as "a five-month odyssey for Garner—to forge an interagency organization from scratch in the Pentagon, deploy it to Kuwait as Coalition forces invaded Iraq, and to follow those forces into the chaos of Iraq to help reestablish order and install a new government." Rudd describes the Phase IV planning efforts as "short, eclectic, incomplete, and lacking in cohesion, although the aggregate planning was more extensive than most believed." Rudd further details how the planning for Phase IV contrasted with the planning for the first three phases, noting, "the invasion of Iraq was elective and deliberate, following military planning that had been conducted for over a decade," Rudd writes. "In contrast, the planning for regime replacement was haphazard and grossly inadequate, with far less time for study and debate."

The decade-long planning process leading up to the invasion of Iraq, as Rudd describes it, was the result of an ongoing battle about the American way of war, and particularly over how and why to deploy American troops. This question led to debates on whether to observe the Powell doctrine of "overwhelming force" or if technological advances meant that wars could be faster and less costly, whether future conflicts would be conventional or asymmetric, and whether US forces should be used strictly for fighting wars or for nation building, as preferred by Rumsfeld.

Central Command engaged in the planning process following Desert Storm, and politics and theoretical debates influenced the final outcome. Operation Plan 1002-90 was the foundation for all the early planning and was used for the initial invasion of Iraq in 1992, and then reworked into 1002-92, which still called for a massive number of US forces. The next iterations, first 1002-94 and next 1003-94, showed a modest directional change under the Clinton administration, which reduced forces by allowing for an increased reliance on coalition forces. The plan further morphed into 1003-98, which allowed for a preemptive strike on Iraq and an increased reliance on technology, and also began considering options related to Phase IV planning in case an occupation or reconstruction became necessary.

General Anthony Zinni, then commander of Central Command, directed his planner, Colonel Steven Kidder, to address these issues, which he did under a planning effort known as "Desert Crossing." For over a year in the late 1990s, Kidder convened interagency partners to develop the basis for a Phase IV plan, but the thirty-six-page document that followed "had never been developed as a plan or even an annex to a plan," according to Rudd. The most that could be shown for progress was a draft annex to 1003-98, which lacked operational detail.

Rudd confirms the SIGIR's description of the Defense Department's best-case-scenario approach toward war planning, which allowed Rumsfeld to choose the most convenient assumptions from the decade-long debate. According to these assumptions, massive troop deployments were no longer necessary, his vision for the Iraq War was one of preemptive invasion not nation building, and Phase IV would largely consist of humanitarian support. The result of these assumptions would provide the foundation for an impossible situation for Garner.

Jay Garner's five months' odyssey began in the Pentagon with "no staff, no workspace, no funding, no knowledge of what planning had been done, and little insight of the situation in Iraq." These limitations should not have been a surprise given Rumsfeld's view of America's limited role in Phase IV operations, but what may have been a surprise was the secretary's constant maneuvers to keep the State Department away from all aspects of the reconstruction effort. In short, Rumsfeld refused to drive the government to support the war effort, which meant that the operational efforts mirrored the highly political, bureaucratic, and uncoordinated characteristics of the previous Phase IV planning effort.

Once in Kuwait, Garner's challenges persisted. He could not develop command relationships or determine which headquarters his agency reported to. ORHA had to build a forward command post in a Kuwaiti hotel and had to beg for access to secure communications while pulling together the pieces of the incomplete Phase IV plans on their personal email accounts. Even when the president solved the turmoil between State and Defense by signing a policy directive delegating authority for the war and reconstruction to Rumsfeld, the challenges remained.

The implications of this failure on ORHA and the occupation effort in Iraq became clear to the world by April 9, 2003, when the transition from Phase III to Phase IV was met with widespread looting throughout the country. Rumsfeld's best-case scenario for reconstruction planning was immediately revealed as wrongheaded, and ORHA had to shift to a focus of "reconstruction and civil administration." Although combat units expected ORHA to move in seamlessly after the invasion, they were unaware of the small organization's lack of a unified plan and resources to comply.

The Iraq phase of Garner's odyssey was brief and would test every assumption made during Phase IV planning. Garner's assumptions were also tested. He understood that he would have sixty to ninety days to establish the interim authority in Iraq, but he was replaced by Paul Bremer and the Coalition Provisional Authority, or CPA, three weeks after his arrival. "Bremer would not only be difficult for Garner," Rudd writes, he was also "planning to reverse many of the initiatives Garner had put in motion."

Bremer's time as the head of CPA not only marked a major shift in US reconstruction policy in Iraq, but it was also marred by two policy decisions that exacerbated the Coalition's inability to maintain order. First through CPA Order Number 1, which the SIGIR describes as "a de-Ba'athification directive that stripped certain former members of Saddam's Ba'ath Party of political influence in Iraq." And second, through CPA Order Number 2, which abolished seven national security organizations, putting half a million Iraqi men immediately out of work. Not only did this mean that Iraq would not be able to contribute to its own security, a capacity that the United States would spend years and billions of dollars to reestablish, but it also loosed countless trained, armed, and unemployed men across the country.

In *My Year in Iraq*, Bremer's memoir about his time at the helm of CPA, he barely touches on the subject of the two orders. Regarding the first order, he largely focuses on the political and media implications and dismisses his detractors by stating that the decision was the president's. His treatment of the second order mostly focuses on all of the measures CPA implemented to mitigate the disasters that his decisions produced, as well as a quote from Jalal Talabani, Iraq's Kurdish leader, that it "was the best decision the Coalition made during our fourteen months in Iraq." Spike Stephenson, the USAID mission director in Iraq during Bremer's time with CPA, was less charitable, referring to the two orders as "spectacular mistakes."

The CPA orders had profound, strategic implications for America's reconstruction efforts in Iraq, but personal dysfunction also continued to hamper the immediate reconstruction effort. Not surprisingly, the dysfunction came down to the normal issues related to the power struggle between the Department of State and the Department of Defense. This tension was concentrated further when Congress initially approved $2.4 billion for the first Iraq Relief and Reconstruction Fund and $18.4 billion for a supplemental fund later in November 2003.

While the funds only increased the infighting between federal agencies, the funding was not done on a whim, strictly speaking. The SIGIR report states that in 2003, "four organizations conducted major reviews of Iraq's infrastructure and economy during the CPA's first three months: Bechtel, USACE, the U.N., and the Center for Strategic and International Studies," each writing reports focused on different aspects of reconstruction and the need for real investment going forward. While these reports would have been valuable during the planning period prior to the invasion, they were only available as the CPA developed their new plan, *Vision for Iraq*, which was followed by an action plan of specific steps that would allow the CPA to facilitate reconstruction.

Vision for Iraq caused problems with CPA partners in Iraq, but the real challenges started when agencies began jockeying for control over the money appropriated by Congress. As the SIGIR notes, "the dispute between the CPA and USAID intensified when USAID discovered that the CPA, a Defense Department organization, would manage all of the IRRF2 funds." USAID submitted an additional budget request to support their reconstruction efforts, but it went nowhere. "These breakdowns, both personal and bureaucratic, would continue to affect the reconstruction program for years to come," the SIGIR notes.

Stephenson describes the details of the turf wars over projects and money in *Losing the Golden Hour*. He cites a meeting with the secretary of state's office where he was warned, "State, having been largely excluded from the planning of the Iraq reconstruction program, was now going to be left holding the bag." This warning was based on speculation and opinion, but the signs of State and USAID's exclusion were more than apparent. Both the Department of State and USAID had also been excluded from CPA's planning process that led to *Vision for Iraq*, and the CPA had created a special Program Management Office, or PMO, which performed a mission identical to USAID.

Through the course of these turf wars, which lasted as long as Bremer controlled the reconstruction funding, Phase IV efforts floundered and failed. According to Stephenson, massive, unrealistic projects were prioritized over smaller, meaningful projects, and contracts were delayed, incurring significant overhead and security costs, all of which ensured progress would remain stalled. "By the time CPA closed its doors at the end of June 2004, the PMO had spent only $366 million of the $18.4 billion IRRF2 appropriation," the SIGIR reports. "Bremer was never able to realize his grand reconstruction vision."

The CPA did accomplish a significant goal while in Iraq by transferring sovereignty back to the Iraqi Interim Government in June 2004, though. This transfer was intended to show progress toward a democratized Iraq, or at the very least, encourage Iraqi leaders to start taking responsibility for their own future. The SIGIR describes another transition that was less important geopolitically, but that may have confirmed State's worst fears about being "left holding the bag," noting, "by the end of June 2004, the Department of Defense formally handed over management of the reconstruction program to the Department of State, and a new set of ad hoc organizations assumed control of a faltering reconstruction mission."

Of course, this "new set of ad hoc organizations" would need a new plan for the reconstruction effort as well. "Operation Plan (OPLAN) Sovereign Iraq" amounted to "the first attempt since the 2003 invasion to develop an interagency strategy for the coordination efforts to stabilize and rebuild Iraq," the SIGIR writes. While the plan would build a foundation for the period following the CPA's control, it continued the same failed practices and policies and ensured the ongoing battle between State and Defense.

As Ambassador John Negroponte took over the management of the reconstruction effort with the new Iraqi Interim Government in June 2004, the divide between State and Defense became even more institutionalized. Where the divide before between USAID and CPA's PMO was largely over who controlled the reconstruction money, National Security Presidential Directive 36 created two new entities representing each agency. The Iraqi Reconstruction and Management Office represented State in the reconstruction effort and the Project and Contracting Officer replaced PMO at Defense. The SIGIR reports, "although NSPD 36 sought to clarify roles and responsibilities, its ambiguities actually aggravated the management situation in Iraq, making it more difficult to achieve unity of effort by fragmenting reconstruction responsibilities."

Even with these ongoing issues, the tone of America's investment in Iraq had changed dramatically with Negroponte's arrival, particularly through the construction of a US embassy in Iraq. The new embassy provided an opportunity for a fresh look at the challenges of reconstruction during a rising insurgency, and it called for new plans to complement OPLAN Sovereign Iraq. While it was different from the haphazard approach that ushered in Phase IV over a year previously, the outcomes were essentially the same, though there was only more at stake.

Robert Earle's *Nights in the Pink Motel* tells the story of how those plans came together. Although in Iraq at Negroponte's request, Earle received his direction, at least initially from Casey, marking an uncommon unity of effort between State and Defense in Iraq. In Earle's telling, this unity filtered all the way down to the mission Casey would give him: writing a successful vision for the counterinsurgency mission in Iraq that brought together military combat, diplomacy, reconstruction, and "all elements of national power," as the president had promised previously. Casey even provided Earle with an idea of what success might look like: "an Iraq built on the principles of representative government, respectful of the rights of its citizens and the rule of law, that it is able to maintain order at home, defend its borders and establish peaceful relations with its neighbors."

In order to plan, Earle and his "red cell" had to review the situation he inherited. This meant coming to terms with the facts on the ground: an insurgency was rising, the presence of the coalition caused at least some of the instability, the battle between Defense and USAID continued, and economic growth was related to security and stability. As Earle developed his plan, he concluded that the effort suffered "from double vision when it comes to the enemy, that we are a long way from coming to terms with the new character of the conflict, and that we are uncoordinated in all aspects of our response—security, economic, and diplomatic/political."

In the assessment of the Negroponte period, the SIGIR report suggests that Earle's conclusion was prophetic. The report covers various themes from the Negroponte period, with sections titled "Contracting Waste," "Case Studies in Waste," and "An Uphill Battle," all providing broad details for what happens when extraordinary resources were present but leaders lacked the ability to coordinate effectively. Iraq was in far worse shape before the invasion than anyone had realized, the CPA's order had set back the security effort at least another decade, and failed systems and infighting ensured that the vision for reconstruction in Iraq would be fraught with fraud, turf wars, and impossible tasks.

The hopeful situation in Iraq that Negroponte inherited in June 2004 was not the same one he passed on to his successor, Ambassador Zalmay Khalilzad, in June 2005. By then, the reconstruction effort had become "a sprawling enterprise," the SIGIR reports, "with separate military and civilian headquarters on opposite sides of the city managing separate—and frequently disconnected—reconstruction missions." Both State and Defense had even undertaken separate strategic assessments to provide Khalilzad a "new strategy" to salvage the effort.

The investment by the US Congress only grew ahead of Khalilzad's arrival, eventually topping $30 billion. Federal agencies still battled for money and projects, but Defense still managed to control about 70 percent of all appropriated dollars. And even so, the SIGIR reports that the outcome was basically the same, that "despite extraordinary outlays of cash and other resources," stability and security had yet to be achieved.

Khalilzad's new plan called for more "jointness" between State, Defense, and other agencies. This meant the political expertise possessed by civilian experts needed to be shared with the military leaders who gained on-the-ground expertise in Iraq's villages and cities every day, and vice versa. Military members were encouraged to attend State Department steering meetings, and representatives from the State Department were invited to attend military assessment meetings. Mostly, the SIGIR notes, it called for a solution for how "the panoply of individual reconstruction projects, stovepiped under various categories, could work better as an integrated whole."

Khalilzad's approach produced real results in areas that had been truly lacking to date. They developed a framework for deploying and assessing the reconstruction efforts that were underway in 2005, and they finally included the Iraqis in the reconstruction process by establishing Provincial Reconstruction Development Councils, which the SIGIR describes as "the first joint Iraqi-American vehicle explicitly aimed at building capacity in regional government institutions." Khalilzad also introduced Provincial Reconstruction Teams, or PRTs, which the SIGIR describes as Khalilzad's "most enduring contribution to the reconstruction program."

The PRT concept involved Khalilzad and Casey both deploying "mixed teams of military and civilian reconstruction personnel in the regions." Although a true innovation, the concept was hampered by the normal failures at the highest levels. It was a State-led program, so Defense did not want to provide security. The departments squabbled over programmatic funding, too, resulting in an immediate "administrative stalemate." Finally, a tortured solution was agreed on, where military members were appointed seemingly at random to fill expert roles in various sections that State could not fill from their own personnel.

The divide between State and Defense was not just seen in the deployment of the PRTs but also in a very public spat between Secretaries Rice and Rumsfeld over a new counterinsurgency strategy called "Clear-Hold-Build." Rice publicly backed the approach in the fall of 2005, but Rumsfeld rejected it. "Clear-Hold-Build" was essentially an admission that the liberation-to-transition approach championed by Rumsfeld had failed. President Bush resolved the dispute in November 2005 by making "Clear-Hold-Build" the centerpiece for his official policy on Iraq, *National Strategy for Victory in Iraq*.

The new effort exemplified by the PRTs on the ground and the unified counterinsurgency strategy at the highest level failed to produce results throughout 2006. The SIGIR writes, "the seeds of this tangled story were planted in 2003, matured in 2004, and finally burst open in 2005." Whatever progress was achieved under Casey and Khalilzad was fleeting, and all the while, Iraq was slipping toward civil war.

With the persistent failures of the *National Strategy for Victory in Iraq*, the president announced yet another strategy that would begin in February 2007. This new strategy doubled down on the counterinsurgency strategy initiated by Casey and Khalilzad and included a surge in military troops for "Clear-Hold-Build" operations, as well as a "civilian surge" that included embedded PRTs, more reconstruction advisors, and greater collaboration with federal partners. As the SIGIR notes, "it would be an 'all-hands-on-deck' approach" that was intended to reverse the trend toward civil war.

Regarding what he calls "embedded reconstruction," the SIGIR gives fairly high marks to the PRTs that were a part of the "civilian surge." They were more integrated with the Iraqis they were trying to serve, they allowed military and civilians to work side by side, and they required that the military acknowledge and accept the reconstruction mission instead of focusing on transition. However, Peter Van Buren reports in his memoir of this period, *We Meant Well*, these efforts continued many of the failures of the recent past at a more local level.

We Meant Well details countless stories of failed projects and millions of dollars wasted. There were off-she-shelf water filtration units that could not purify the salty water and wound up unused, broken, or stolen; a milk collection center that was funded and built but that never operated; and a $2.58 million chicken processing plant that only processed chickens when high-level visitors arranged a visit. Van Buren tells numerous stories like these, suggesting that even the commitments made during the announcement of the surge strategy would not be enough to reverse the Coalition's fortunes in Iraq.

Although generally positive on the progress represented by the idea of the troop surge, the SIGIR's assessment is in fact more in line with Van Buren's. Although the two new leaders, General David Petraeus and Ambassador Ryan Crocker, emphasized jointness in their approach, "a truly joint command structure never evolved," USAID remained largely excluded, and reconstruction continued to fail. "The Joint Campaign Plan that Crocker and Petraeus promulgated was predicated on the belief that stopping openly violent clashes would allow Iraqis to gradually resolute their conflicts through elections rather than militias," the SIGIR notes. "Such a transformation was both elusive and extraordinarily costly—for Americans and Iraqis—in lives and national treasure."

In fact, all aspects of Phase IV resulted in elusive success at an extraordinarily high cost, but this was not due to the fact that the Bush administration lacked a plan as the Phase IV narrative suggests. Rather, there were many plans, too many perhaps, and each new period of the Phase IV effort came with updates or entirely new approaches intended to salvage the nation's investment in the war effort to date. What was lacking was not a plan, but rather leadership to overcome the natural tendencies of bureaucracies when money and legacies are involved.

5

COIN Toss

The Surge Narrative

THE SURGE narrative is a prominent narrative of the Iraq War, and to a lesser extent, the war in Afghanistan. As it pertains to the Iraq War, the surge narrative essentially states that General George Casey had basically lost the Iraq War by the end of 2006 due to his ineffective strategy, leadership, and execution, but in February 2007 General David Petraeus took command and led the nation to victory. This narrative has become popular in many circles, especially military circles, and often bleeds into discussions of who is to blame for the ultimate failure to stabilize Iraq after the invasion, who is to blame for the rise of the Islamic State and Iran in the region, and how American military force should be used in the future.

The surge narrative relies on the basic facts of the Iraq War to make its case: the situation in Iraq was deteriorating throughout 2006, President Bush decided to surge additional troops and change commanders in early 2007, and some important successes followed those changes. However, those who subscribe to the surge narrative have to order and present their facts in just the right way to support their case. When these same facts are removed from the surge narrative, a far more complex story is revealed.

The foundation of the surge narrative can be found in the challenge the Iraq War presented to America's role in the world. In postmodern wars, all of the traditional rules of combat are thrown out the window: front lines are replaced by amorphous battle space, insurgent forces mix with the civilian community, and conventional military strengths become tactical weaknesses. Perhaps most frustrating for superpowers in postmodern wars is the fact clear and decisive victories are difficult to achieve.

America's frustration with these aspects of counterinsurgency warfare, or COIN, came to a head in Iraq in 2006, and Afghanistan was not far behind. Public opinion of the effort dropped tremendously, accomplishing the broad goals of the Bush administration proved difficult, and a civil war between Iraq's Sunni and Shiite populations was underway. In 2006, America needed a hero who could tell the Iraq story it wanted to believe, and General David Petraeus was ready to be that hero.

Many authors and commentators have embraced the surge narrative, some by simply accepting it as fundamental fact. Others seem keenly interested in preserving and even promoting it, namely, Peter Mansoor's memoir, *Surge: My Journey with General David Petraeus and the Remaking of the Iraq War* and Kim Kagan's *The Surge: A Military History.* Both Mansoor and Kagan tell the story of the war after President Bush's decision to surge troops in very different ways; however, both authors clearly embrace the narrative.

For the surge narrative to work, it must convincingly present a binary proposition. On one hand, everything that happened before the surge was a failure carried out by incompetent generals, and on the other, only David Petraeus was capable of turning the situation around. While both authors assert or assume that Petraeus's execution of the surge resulted in a stunning strategic success, which is the true centerpiece of the surge narrative, Mansoor, in *Surge*, goes out of his way to make the case that the generals who preceded Petraeus were simply inadequate. However, this common assertion proposes at least two significant problems.

First, General Petraeus was one of the generals in charge of the war prior to his arrival with surge forces in 2007. After commanding the 101st Airborne Division in 2003, Petraeus led the Multi-National Security Transition Command-Iraq (MNSTC-I) organization in 2004 and 2005, which was responsible for training and equipping Iraqi security forces. If excluding General Petraeus from the generals in charge of the early war effort was not bad enough, surge narrative proponents often also ignore the fact that the forces Petraeus trained under MNSTC-I would become a part of the civil war that would have to be quelled later as well.

Second, there were in fact significant successes in Iraq prior to the surge, successes that run counter to the proposition offered by the surge narrative. One was the Sunni Awakening, where Sunni sheikhs especially in Iraq's Al-Anbar province began to work with the coalition forces to defeat Al-Qaeda, a program that began in 2006 under General Casey. To deal with this inconvenience, Mansoor gives a nod to those who implemented the strategy, but he states that only Petraeus could see its true value when he took command. Mansoor also points out that the idea for the Awakening came from Petraeus when he was a division commander in Iraq at the beginning of the war, at least suggesting that one of Casey's greatest achievements actually belonged to Petraeus.

The extension of the basic proposition of the surge narrative to authors like Mansoor is the requirement that one must either be on the team that promotes the successes of Petraeus or they must be against it. According to Mansoor, when Senator Harry Reid dismissed the surge he did it for political reasons, but when Senator John McCain toured Iraq and doubled down on the surge as a part of his presidential campaign, he was an informed leader. When Admiral William "Fox" Fallon, Petraeus's boss during his time as commander in Iraq, sought to review conditions in Iraq and the resources Petraeus required, Fallon was second-guessing and undermining the commander in the field. However, there is no similar condemnation for Jack Keane's efforts as a retired four-star general, who worked through a civilian think tank and political back channels to second-guess and undermine George Casey, likely because Keane was promoting the surge.

Unlike Mansoor, Kagan appears less interested in making a case for Petraeus and his successes through the surge. Instead of focusing on the work Petraeus and Ambassador Ryan Crocker pursued through the civil-military relationship, Kagan focuses on General Ray Odierno's coordination of brigades and divisions on the Iraqi battlefields at the operational level. Kagan attempts to preserve the surge as military history, focusing on which units attacked which enemy forces, where, and when.

Kagan's *The Surge* takes a chronological look at the various operations that made up the surge. Examining the war from the battalion level, she shows how the first four years of war in Iraq set the foundation for what would follow in the surge, chronicling the efforts to combat terrorists in the Baghdad Belts, to bring Iraqis into the fight, and to otherwise bring peace and stability. But even with this lack of overt and unquestioned promotion of the surge as a strategy, Kagan accomplishes many of the same goals as Mansoor.

Most notably is Kagan's view that the surge was ultimately extremely successful, which is a fundamental assumption of her work. She notes, "the general stabilization of central and western Iraq has allowed both U.S. and Iraqi leaders to focus on problem areas in the south and north that had been under-resourced for a long time." While this may have been the case for a time in 2008 when writing her book, stability did not last.

Mansoor, whose 2013 memoir had the benefit of more time to examine the effectiveness of the surge strategy than Kagan's, is equally committed to declaring it a success. In a chapter devoted to Petraeus's congressional testimony in September 2007, Mansoor cites remarks from Senator John McCain, writing that a "premature U.S. withdrawal from Iraq would be a perceived victory for the jihadists, and it would embolden Iran and lead in due course to the creation of a terrorist sanctuary in the heart of the Middle East." While this statement is used to support the surge in Mansoor's 2013 memoir, by 2014 it was clear that the low-level civil war between Iraq's Sunni and Shia that surge was supposed to defeat had expanded as the Islamic State and Iran have gained regional power.

Mansoor, Kagan, and others seem committed to preserving the argument of the surge as an effective strategy. Linda Robinson, Tom Ricks, Paula Broadwell, and Fred Kaplan all explore the wars in Iraq and Afghanistan, or Petraeus, through the surge narrative, often just as forcefully as Mansoor. While these works tend to accept the basic proposition of the surge narrative, they also reject the conclusion that the surge was a tremendous success.

Linda Robinson relies on the surge narrative for her book *Tell Me How This Ends.* She notes that the situation in Iraq was dire, the administration had no endgame from the start, military leaders were ineffective in standing up to their civilian leadership, they were equally ineffective in coordinating with their civilian counterparts at the Department of State, and by 2006, the "war was heading over the cliff." However, she writes, "in a war of mishaps and missteps, for once the right person was in the right place at the right time."

In Robinson's telling, when Petraeus was finally placed in command of the war effort and was allowed to implement the surge, the war effort went from losing to winning. Part of the solution was Petraeus himself. He had the right intellect, the right temperament, and the drive to see the effort through. But another part of his success was the team he had assembled, a group of the greatest thinkers in the military, Rhodes Scholars, PhDs, and partnerships with think tanks, media representatives, and others who could help Petraeus turn around the struggling war effort. As her title implies, Robinson is not fully committed to declaring the surge a strategic success, but her reliance on the surge narrative at least suggests that if the war effort were to turn around, it could only be done by someone like Petraeus.

After Robinson, the surge narrative began to gain traction, not just as applied to the Iraq War, but also to Petraeus's later work in Afghanistan. Tom Ricks again relies on the narrative to Iraq in *The Gamble: General David Petraeus and the American Military Adventure in Iraq, 2006–2008*, and Paula Broadwell examines Petraeus's leadership in Afghanistan in *All In: The Education of General David Petraeus.* Fred Kaplan attempts to dig deeper into both by applying the narrative in *The Insurgents: David Petraeus and the Plot to Change the American Way of War.*

Ricks aggressively asserts the surge narrative in *The Gamble*, explaining the Iraq War in far darker terms than Robinson, while sparing Petraeus. By way of showing his acceptance of many of the assumptions that Robinson goes to great pains to establish in her early chapters—that the war was essentially a dead end and that Petraeus's predecessors had squandered any chance to win—Ricks simply assumes their veracity as the basis for his story. He gets to the heart of the surge narrative in his first chapter, stating, "the answer for what to do in Iraq would come largely through one person, Gen. David Petraeus, who over the next year would lead the way in determining how to revamp the U.S. approach to the war." While this puts the war effort in easily understandable terms, it does not account for the extraordinary military, political, and strategic complexities of the war in its early years.

If Ricks's establishes the surge narrative for Iraq by attempting to show how all of his successors failed, Paula Broadwell's *All In* is its opposite in some respects. First, Broadwell's book applies the surge narrative to Petraeus's time in Afghanistan. Second, she chooses to focus more on Petraeus's successes and how he earned them and less on the incompetence of his predecessors. But even with these differences, the result is essentially the same.

Although *All In* has been reassessed in recent years, it is still an important tool for examining how far the surge narrative had gone, especially at the time of its publication. In extending the surge narrative to Afghanistan, Broadwell does not stray far from the basic storyline. In Broadwell's telling, Petraeus had been an unmitigated success in Iraq in using the additional troops to pacify Iraq, and he would attempt to do the same for the troubled Afghanistan war effort.

To Broadwell, Petraeus's force of will alone was "a new strategic force loosed in Kabul"; he was a guy who "would never give up"; and in the end, he was a leader who "left an indelible mark on the next generation of military leaders as a role model of a soldier-scholar-statesman." When she is not describing Petraeus's personal greatness, she is comparing him to Ulysses S. Grant, or focusing on his five-mile run time, or his eighteen-hour days. She also quotes his military disciples, people whose legacies are inextricably linked to his, quoting numerous starry-eyed email exchanges.

Where Ricks and Robinson seem dedicated to portraying Petraeus as the right man at just the right time in Iraq, Broadwell shows his success to be his destiny by construing every aspect of his life as a part of the education that would prepare him for Iraq and Afghanistan. Although he had never been to combat before Iraq, Petraeus spent four months in Haiti, which would serve as a "primer for a similar challenge he would face on his second tour in Iraq, when he had to create a new command to recruit, train, equip, advise and build infrastructure for the Iraqi military and police." He worked with special operations in Bosnia, too, which would serve "him well later in Iraq." Iraq and Afghanistan would also prepare him for his following assignment, too, as Broadwell writes, "no one operated at the strategic political-military level better than Petraeus, and heading CENTCOM was the ultimate pol-mil job."

Broadwell does address some of the criticisms that have been leveled at Petraeus throughout his career, but only to dismiss them. What some describe as Petraeus's "endless campaign of self-promotion" is discounted as envy and jealousy. When media criticism arose of Petraeus's switch from counterinsurgency in Iraq to the more lethal counterterrorism approach he used in Afghanistan, Broadwell notes that "the nuance was lost on many journalists." Other critics are "surge skeptics and revisionist." When Petraeus was asked to lead the mission to train Iraq's military and police, a mission that some would claim was a failure, Broadwell blames the shortfall on his superiors who did not provide him the staffing he needed to accomplish the mission, adding, "Petraeus reacted the only way he knew how—by putting his shoulder to the wheel and pushing forward as hard as he could, all day, every day."

Where Ricks propels the narrative by tearing down Petraeus's peers and Broadwell reinforces it through her praise of Petraeus, Fred Kaplan employees both in his book *The Insurgents*. Kaplan claims in the acknowledgments to not be "in the business of hagiography," and he even recognizes that the surge narrative exists, stating, "in the euphoria over the undeniable improvements in Iraq, the emergence of near-normal everyday life, a tendency arose both in the field and back in Washington to give all credit to the American Military, to the surge, or to General Petraeus and his counterinsurgency strategy." Even with this admission, however, he still proceeds to embrace the narrative.

Like Ricks, Kaplan portrays Petraeus as the only leader who understood that the nation was involved in a counterinsurgency battle, especially when contrasted with his peers. "Casey," Kaplan writes, had "never been in combat and had scant experience of any sort in the Middle East." In fact, he had never "been interested in any kind of war except the big kind." To support this assertion, Kaplan adds that Casey had "even turned down an offer to join Delta Force." Although Kaplan acknowledges that Casey had been given "contradictory mandates" from Rumsfeld, and that the first thing he asked when he entered his Baghdad headquarters was, "where is my counterinsurgency plan," Kaplan paints Casey as the reason the war effort in Iraq failed.

Like Broadwell, Kaplan also perpetuates the surge narrative by highlighting the general's greatness. Unlike the others, though, Kaplan does take several opportunities to point out Petraeus's outsized ambition, his willingness to give glowing self-assessments, and his willingness to circumvent his chain of command to advance his own career. But even so, Kaplan bases his book on the idea that Petraeus was probably, in fact, the greatest general of his generation as seen by his successes in Iraq and Afghanistan, the kind of guy who survived being shot, beat prostate cancer, a parachute accident, and all of that was before he became "the miracle worker of Iraq."

But even with the help of a "miracle worker," Robinson, Ricks, Broadwell, and Kaplan all state Petraeus did not achieve success and that the surge was a failed strategy. If the wars were disastrous before the surge, Petraeus only provided hope for unrealized improvement. Their backtracking shows the central contradiction of the surge narrative, and it also undermines Mansoor, Kagan, and others who build their cases around the successful outcomes of the surge.

Robinson, writing early in the story of the surge, provides a broad assessment of the effectiveness of the surge. "The government still had not managed to take any major steps on national reconciliation, and it was to this goal that all efforts were now bent," she writes. Without this reconciliation, "the gains bought at such a high price would not last." Ricks agrees in *The Gamble*, writing, "in revising the U.S. approach to the Iraq war, Petraeus found tactical success—that is, improved security—but not the clear political breakthrough that would have meant unambiguous strategic success." Ricks takes this a step further, adding, "at the end of the surge, the fundamental political problems facing Iraq were the same ones as when it began."

Assessing Petraeus's successes in Afghanistan, Broadwell explains the lack of strategic success in terms that are favorable to Petraeus, choosing instead to paint the situation in personal terms that conveyed Petraeus's commitment to the war effort. According to her, the general truly wanted to achieve decisive victory in Afghanistan, presumably like he had done in Iraq, but he knew that he had yet to achieve such success. As he departed Afghanistan, he tried "to put various thoughts behind him and compartmentalize his feelings," adding that he answered important correspondence on his last flight out of country.

Presumably in line with his position against hagiography, Kaplan admits the failure of the surge as well. "It was unclear whether the Iraqi government had really adopted the same goals as its American rescuers, whether the country's sectarian factions were capable of reconciling with one another, or whether Prime Minister Nouri al-Maliki and his top ministers were interested in reconciliation," he writes. "Nor was it clear," he adds, "even if all did end well, whether the crushing costs—in lives, money, and strategic position, of Iraq and the United States—had been commensurate with the gains."

While the surge narrative has remained central to the broad understanding of the war in Iraq, some have provided counter arguments. General George Casey's brief memoir, *Strategic Reflections: Operation Iraqi Freedom, July 2004–February 2007*, Colonel Gian Gentile's *Wrong Turn: America's Deadly Embrace of Counterinsurgency*, and Douglas Porch's *Counterinsurgency: Exposing the Myths of the New Way of War*, all provide details and arguments that counter the surge narrative.

In *Strategic Reflections*, Casey provides insights into the strategic context that set the conditions for him in Iraq. The surge narrative criticizes Casey for not being interested in applying counterinsurgency strategies and tactics to the Iraq War, and while not addressing the narrative directly, Casey states, "we laid out a counterinsurgency strategy that sought to use the full spectrum of military and civilian tools to separate insurgents and extremists from the Iraqi people and defeat the insurgency while we restored Iraqi capacity to govern and secure the country." His plan was to "neutralize the insurgency in the Sunni Triangle," "secure Baghdad," "block the borders of Iraq to disrupt the flow of support to the insurgency," "assist in building the ISF," and "sustain support for coalition force efforts in Shia and Kurdish areas."

Casey describes how some of his commanders were having success at the tactical level, including the Sons of Iraq effort as a part of the Sunni Awakening, which Petraeus is often credited with. "Sunni engagement was something that General Abizaid and I had been pushing for some time," Casey writes. "We had pressed hard for an engagement strategy with the Sunni population as a means of driving a wedge between the general Sunni population and the terrorists and insurgents." Casey achieved considerable success through this effort, stating that it "marked a major turning point not only against al Qaeda, but with the population of Anbar Province."

Porch and Gentile take on the surge narrative directly, providing historical analysis of how counterinsurgency has become an ideology of its partisans. Both works examine counterinsurgency efforts decades back or longer and point out the weaknesses of the movement's claims. Both works are also damning of those who promote the surge narrative, with Gentile's assessment more general and Porch's more pointed.

Detailing counterinsurgencies back to the middle of the nineteenth century, Porch examines many challenging aspects of the prevailing enthusiasm for counterinsurgency, focusing on the dishonest and manipulative use of history, analysis, and narratives by movement proponents to achieve their goals. Porch writes: champions of counterinsurgency "concoct a wartime story line of official optimism, claim progress and success with spinfests fueled by an annual $4.7 billion annual Pentagon PR budget that allows contracted media mercenaries attached to military staffs to maintain a confident narrative of counterinsurgency momentum, expand in-country goals, coax more men and resources out of skeptical governments, keep civilian policymakers off balance, and torpedo their military rivals with accusations that they 'don't get COIN.'" Strategic miscalculations during wartime created "the opening for 'new school generals' like David Petraeus and Stanley McChrystal with a timely assist by John Nagl's 2005 *Learning to Eat Soup with a Knife* fantasy historical account of British COIN mastery to reassert a COIN narrative in hibernation since Vietnam." Porch concludes that the goal of such efforts is "to discredit conventional warfare rivals, snag the top echelons of the US Army, and proceed to hijack the policy."

Porch identifies Ricks's *The Gamble* as establishing "the COIN-dinista narrative of how General David Petraeus and a cohort of defense intellectuals, both in and out of uniform, designed the 'surge' to rescue a failed coalition strategy in Iraq." The key elements of Ricks's narrative are the deterioration of the situation in Iraq toward civil war in the summer of 2006, which created pressure on President Bush to shift his strategy. "But help was in the wings," Porch writes, "in the form of a team of maverick officers and their dynamic leader, General David Petraeus in the US Army's TRADOC in Kansas, where the general had gone after his first Iraq tour." To Porch, this is not just Ricks's convenient way of framing the latter part of the Iraq War, but it continues in a tradition of "journalists who have long made common cause with colonial soldiers in the battle for the public mind at home, with an identifiable cast of heroes and villains, and a happy ending."

At the heart of the narrative, according to Porch, is a field manual; Petraeus supervised the development of new counterinsurgency doctrine while stationed at the US Army's TRADOC in Kansas, *FM 3-24: Counterinsurgency*. When the conventional approach failed, Porch writes, "the 'Petraeus guys' rode to the rescue with a self-reverential doctrine of a renewed American way of war for the twenty-first century," allowing them to "rebrand themselves as the White Knights of American Exceptionalism." If *FM 3-24*, is central to this narrative it is also its central weakness, Porch points out, and for a number of reasons.

First is that the doctrine developed in *FM 3-24* did not produce the political and strategic goal of a political solution in Iraq, much like some proponents of the surge narrative admit. Second, Porch notes that *FM 3-24* "failed to supply a blueprint for strategic reversal in Iraq beyond a compendium of predictable platitudes about requirements for 'legitimacy' and 'effective governance' lifted from British theory 'with adjustments for contemporary context.'" And third, *FM 3-24* was released only months before Petraeus took command in Iraq, arriving at least eighteen months after American forces had already adapted to the insurgency. The real purpose of *FM 3-24*, Porch argues, was "to legitimize Petraeus' cult of personality by discrediting the effort of 'conventional' predecessors—indeed rejecting the notion that conventional war will play any part in the United States' imperial mission or in national security *tout court*."

In a more general way, Gentile provides historical analyses of what he calls the "surge triumph narrative" as it pertains to common understanding of the counterinsurgency efforts in Malaya, Vietnam, Iraq, and Afghanistan. He contends that the narratives for these wars follow a single arc: a counterinsurgency war looks to be on the verge of defeat and the proposition of a "better war" emerges, one that is led in a drastically different way, following classic counterinsurgency methods, and led by a smarter general who usually wins the war only to have it lost by his weak-willed political masters. The problem, Gentile points out, is that these narratives find very little historical support.

One of the key parts of the narrative that Gentile debunks for all four of his wars also supports Casey's modest suggestions that he indeed had a counterinsurgency strategy. Gentile argues that purveyors of these narratives often argue that the savior general implements "a very different doctrine" and takes the war in a "radically new direction" once in command. Gentile goes to great lengths to show how commanders in Iraq who would later argue that Petraeus did in fact bring in a strategic and tactical sea change with him were on the record describing their own implementation of successful counterinsurgency operations in Iraq under Casey.

Where Casey merely mentions that the Sons of Iraq program was underway during his tenure, Gentile takes it a step further. "The surge of troops led be General Petraeus was not the primary cause for the lowering of violence that began in the early fall of 2007," he writes. There were instead a "combination of critical factors" that led to the peace, chief among them was the Sons of Iraq effort started under Casey.

Casey points out another success that contributed to achieving his counterinsurgency goals: the national elections in Iraq. His efforts in this area do not just fly in the face of his critics' claims, they also sound remarkably similar to actions for which Petraeus received praise. Petraeus held successful elections when he was a commander in Mosul, but Casey held two national elections that were crucial to drive progress in Iraq. Casey writes of the 2005 elections, "on January 30, our detailed preparations paid off as over 8 million Iraqis—58 percent of the eligible population—turned out to vote." It was an enormous task, one that a few months prior the president had described as Casey's "most important" effort. Less than a year later, on December 15, Casey would hold a national election again, this time with turnout of 76 percent.

While Casey takes pride in the success of these elections, his detractors argue that the elections actually led to national division due to the low Sunni turnout. They argue the failure of demographic representation at the polls would eventually lead to the civil war in Iraq, but the elections were not the only precursor. Another factor, which proponents of the surge are silent on, could be attributed to Petraeus. By February 2006, with the civil war officially underway, Shia militias were fully integrated into the police forces trained under Petraeus's command, thus providing legitimacy for the death squad executions regularly conducted in Sunni neighborhoods.

Casey goes to great lengths to point out the importance of the MNSTC-I mission while he was commander in Iraq. He notes that his mission from the beginning was to "hand over authority to a sovereign Iraqi government," and a crucial part of doing so was the acceleration of the "program for training Iraqi security forces with an eventual goal of an Iraqi army of 27 battalions and an overall ISF number." Key to achieving that mission was his partnership with Petraeus, then the MNSTC-I commander.

Although handing over responsibility for security to the Iraqi security forces was President Bush's stated mission, Casey continually notes that the mission was simply not being met. In fact, having the "Police in the lead" with an Iraqi Army in support were listed as parts of the keys to Casey's counterinsurgency effort, but routine assessments showed that "police and border guard capacity was still particularly weak and needed to be improved," and that MNSTC-I's efforts were not proceeding as quickly as necessary. Casey does not criticize Petraeus directly, but the implications are clear: the general who would later be seen as Iraq's savior failed to achieve the president's stated mission, and this failure contributed to the civil war.

Gentile challenges the claim that the surge was successful because it quelled violence in Iraq by pointing out that while the Iraqi military was not gaining capacity under MNSTC-I, the death squads that were formed within the Iraqi national police system were becoming increasingly efficient. From February through December 2006, national police death squads pursued an objective of cleansing key Baghdad neighborhoods of Sunni residents, often with retaliation by Sunni suicide attacks in Shia neighborhoods. By December 2006, the death squads embedded in the national police had largely achieved their goal through murder and intimidation, Gentile argues, and it was their achievement of their sectarian goals that served as a critical factor in the decrease of violence in 2007, and not the surge that would start a few months later.

Through his examination of counterinsurgency operations in terms of strategic communications, a practice that invariably involves the use of narratives, Porch captures what is probably the most significant argument against the surge narrative. He writes that contemporary proponents of counterinsurgency "are basically romantics, whose strategic communications target politics and society at home while they profess to apply paternalistic theories onto biddable populations required to show gratitude for their improved conditions by outing pistoleros, bandits, terrorists, and insurgents in their midst." In *Surge*, Mansoor cites General Casey's warning that the surge would result in a "localized, temporary improvement in security," in order to discount it by showing all the ways the surge was successful. Mansoor's assertion presents an important question about the surge narrative: if the situation in Iraq did in fact falter after Petraeus and his increased troop numbers departed, as Robinson, Ricks, Broadwell, Kaplan, and many others have noted, then how could the surge be seen as the enormous success its proponents claim? In Porch's terms, the answer may be that the surge was not a strategic military success at all, but rather a domestic political success that provided the Bush administration with the necessary narrative to finally declare a long elusive victory.

6

Heroes and Victims

The Hero and Victim Narratives

AMERICA'S understanding of veterans in recent times can be reduced to two basic narratives: the hero narrative and the victim narrative. The former narrative states that those who have served in uniform are heroes and due eternal gratitude, while the latter narrative asserts that those who have served are victims and are due eternal support. These narratives accommodate various social and political needs: they serve as shorthand, they affirm political preconceptions, but they often fail to match the same when described by those who experienced them.

These narratives represent America's divided political culture. The hero narrative is the dominant narrative of the political right, an ideology that promotes a strong national defense and appreciates military service to promote it. The victim narrative, conversely, is largely the view of the political left, factions of which have historically critiqued the nation's wars and military service. Both sides of the political spectrum have also been willing to appropriate both narratives to manipulate the public for political gain.

Although these narratives can be seen across American culture, they have arguably been most pronounced within country music. Mainstream country music has increasingly sought to appeal to conservatives in recent years, and the wars in Iraq and Afghanistan provided an opportunity for musicians to explore conservative values during wartime. Alternative country music, which fills a far narrower niche within the country music audience, has responded with protest music that appeals to the genre's more urban and liberal audience.

Although the list of mainstream country music songs extolling the heroism of America's veterans is lengthy, no artist has taken on the message like Toby Keith, especially in his song "American Soldier." The song tells the story of a man defined by his sense of duty, commitment to his nation, and willingness to sacrifice for both, even to the point of death. As the title suggests, he is a representative for all who serve, and he is a hero.

There are numerous alternative country songs that speak to the costs of the wars in Iraq and Afghanistan, but "Mama Bake a Pie, Daddy Kill a Chicken" by the Drive-By Truckers provides a poignant example. Like Keith's song, it also tells the story of an American soldier, but their young character is returning home after losing his legs. In a broad sense, the band is singing about America's class structure, but their specific story addresses the themes of loneliness, loss, and shame, and these characteristics of victimhood are presented as continuations of the character's war.

These two songs represent the larger discussion happening within the divided parts of American culture. These narratives are reflected in books about the wars, news coverage, and movies. Although some social and political commentators have challenged these narratives in recent years, they did not become a part of the cultural conversation in any significant way until 2014.

In June 2014, army captain Benjamin Summers wrote a piece for the *Washington Post* titled "Hero Worship of the Military Is Getting in the Way of Good Policy." Summers acknowledges that some veterans deserve praise for heroism, but identifying all who served as heroes has numerous unintended consequences: it diminishes the sacrifices made by actual heroes and it frames the policy debate for wars, funding, and other resources around either being for or against the troops.

While music directed to a liberal audience highlights the victim narrative, liberal political commentators have taken the lead in challenging the hero narrative. Authors Cara Hoffman and David Masciotra address the problems with the hero narrative in the progressive online magazine Salon.com in July and November 2014, respectively. In "Stop Calling Soldiers 'Heroes': It Stops Us from Seeing Them as Human—and Dismisses Their Experiences," Hoffman calls for specific societal changes to address issues like veteran suicide. Masciotra's article, "You Don't Protect My Freedom: Our Childish Insistence on Calling Soldiers Heroes Deadens Real Democracy," aggressively challenges the hero narrative from a populist perspective.

Hoffman and Masciotra address similar challenges facing veterans, particularly of the wars in Iraq and Afghanistan. Both analyze the hero narrative in literary terms, Hoffman calling for an end to attaching "cowboy-and-Indian narratives" to the "sad and extremely common reality of violent conflict," while Masciotra reduces the reliance on the hero narrative to a simplistic need for a villain. Interestingly, both Hoffman and Masciotra attack the hero narrative while embracing the victim narrative. Hoffman writes that labeling veterans as heroes merely turns them into fictional characters, which is "little consolation for the weight of physical and psychological injury these people incur." Masciotra writes, "far from being the heroes of recent wars, American troops are among their victims."

Masciotra's article has much in common with Rory Fanning's October 2014 piece on the progressive AlterNet.org titled, "Army Veteran: Why Do We Keep Thanking the Troops for Their 'Valor'?" Fanning's piece considers a concert dedicated to veterans, and why icons of the left like Bruce Springsteen are so willing to support the effort. Finding political common ground with both Hoffman and Masciotra, Fanning concludes his piece by arguing that it is "years past time when anyone here should be able to pretend that our 18-year-olds are going off to kill and die for good reason."

The victim narrative also became the subject of critique in 2014. While not always coming from explicitly conservative writers and sources, critiques of the victim narrative have been mainly from the conservative worldview. That is, at the core of the victim narrative critique are notions of self-reliance, personal responsibility, and service and sacrifice, all of which are characteristics idealized within the conservative worldview.

Prior to his service as the secretary of defense in the Donald Trump administration, retired marine general James Mattis directly addressed the subject of victimhood in a speech at San Francisco's Marines' Memorial Club in April 2014. "While victimhood in America is exalted I don't think our veterans should join those ranks," Mattis said. While many discuss posttraumatic stress disorder, Mattis argued instead for the concept of posttraumatic growth, where individuals have seen personal growth from their experiences.

Mackubin Thomas Owens relies on Mattis's speech for his article "Life after Wartime: Combating the Veteran-as-Victim Narrative," published in the conservative magazine the *Weekly Standard* in June 2014. Like both

Hoffman and Masciotra, Owens invokes the literary definition of the hero, but unlike them, he argues that veterans should be treated as heroes and as objects "of admiration and respect." He agrees that growth can happen after trauma, and often does, and asserts that the victimization of veterans is largely a construct of liberal media and academia.

While not as clearly conservative as Owens's argument, Starbucks CEO Howard Schultz and Rajiv Chandrasekaran write from a similar perspective in their October 2014 piece in the *Washington Post* titled "Want to Help Veterans? Stop Pitying Them." The authors published the piece with the release of their book, *For Love of Country: What Our Veterans Can Teach Us about Citizenship, Heroism, and Sacrifice*, the title of which clearly indicates their perspective. They try to navigate between the hero and victim narratives, arguing that a "better recognition of the overall veteran experience—the bad, the good and everything in between—is essential to forging a lasting compact between those who have served and the rest of us."

Nearly all authors on either side of the issue accept that a key reason for the development of these narratives is the growing distance between the nation and its military. That is, these narratives arise out of a lack of common experience, but this distance does not have to be permanent. Hoffman writes that the broader public can gain a stronger understanding of these experiences "by listening and by opening our eyes," she writes, "by letting people who are returning take off the masks society insists they wear."

Fortunately, the wars in Iraq and Afghanistan have been presented as the opportunity to listen to the broader public in an unprecedented manner, thanks in a small part to embedded reporters, combat camera units, and camera ready and internet capable soldiers who have provided more information from the front. Perhaps most interesting, though, is the growth of oral history efforts for veterans, which aim to capture first-person combat and wartime narratives. Recent collections consist of many hours of unedited, in-depth interviews with veterans, often relying on a scripted set of questions and resulting in recordings or transcriptions of personal accounts, all of which can be used by academics and writers to attempt to analyze the veteran experience from the current wars. Several books in recent years have capitalized on the opportunity provided by the growth of this format, namely, Larry Minear's 2010 book, *Through Veterans' Eyes: The Iraq and Afghanistan Experience*; Lori Holyfield's 2011 book, *Veterans' Journey Home: Life after Afghanistan and Iraq*; and Trish Wood's 2006 book, *What Was Asked of Us: An Oral History of the Iraq War by the Soldiers Who Fought It.*

While there is great opportunity presented in new oral histories, there are also challenges, depending on how the authors and editors choose to present and interpret the interviews. Minear and Holyfield, for instance, approach their work by editing various interviews, news sources, and other commentary into similar passages and categories while adding their own thoughts. Conversely, Wood engages in almost no editorializing, choosing rather to simply present the words of the veterans. These approaches lead to vastly different interpretations, with Minear and Holyfield relying on the victim narrative to unify the themes and messages of their works, while Wood allows the interviewees' words to speak for themselves.

Minear's approach in *Through Veterans' Eyes* seems aimed at confirming his opinion from the outset. Although he claims to only want to make his book one "in which veterans tell their own stories, with [his] own role limited to setting the stage and filling in a few blanks," he does not succeed, seeming instead to selectively edit the interviews and inject his own conclusions. Too often, the wars are not presented through the eyes of veterans, as his title suggests, but rather through the lens of victims, where nearly every grievance is aired as fact and every hardship and reality of warfare is determined to be yet another example of war's theft of the soldier's humanity.

Minear begins his analysis of veterans as victims immediately in *Through Veterans' Eyes* by dividing the book into three sections, "The Setting," "The Experience," and "Reentry." In "The Setting," he tries to grasp the unique complexities of the broader wars through the words of the interviewees. It is not just the geography that he is attempting to describe, but also those who are involved, concluding that they are different: "they belong to an all-volunteer military," he writes, many are "'citizen-soldier' members of the National Guard," and many more than before are women.

Minear describes the equally unique wars that these soldiers endure in "The Experience." It begins with a discussion of the politics and beliefs that led many to join during wartime and how those politics and beliefs conflict with their professionalism. He focuses a chapter on "coping with unfamiliarity and violence," concluding somewhat obviously that "most of those interviewed found soldiering in Afghanistan and Iraq to one degree or another unnerving and disorienting." They faced difficult ethical issues in combat, they were "operating under duress," they faced unwinnable circumstances fighting a counterinsurgency war among a largely civilian population, and they had a strange and strained relationship with the media. Many might describe these as typical wartime experiences.

In "Reentry," Minear discusses the veteran's reintegration into civilian society. He goes to great lengths to discuss the experiences each service member had, state and federal programs that have been effective and ineffective, and the relationships between different generations of veterans. The discussion of how a civilian nation addresses the needs of those it sent to war is an important one, but focusing on the government services provided to veterans suggests that Minear is at least working from the perspective that all who served are broken or wounded, that they are victims of these wars.

Linda Holyfield differs from Minear's approach in *Veterans' Journeys Home*, but the outcome is the same. She is the daughter of a veteran and a sociology professor at the University of Arkansas, and both facts inform her writing. But no matter how different her approach, she still writes in a way that is condescending to her subjects. "As sociologists," she writes, "we are in the business of examining how social processes work in order to understand who benefits and who loses from structural arrangements." If one were to judge her on how well she accomplishes this task in *Veterans' Journey Home*, he/she might conclude that her results were mixed at best, due in large part to the enormous difference in the way she approaches the first and second halves of her book.

The first half of Holyfield's book essentially makes the argument that contemporary social and cultural conditions combine with military structures and practices and postmodern military combat to create an epidemic of traumatic injury for a generation of veterans. Specifically, she argues that the nature of today's service members, the military socialization process they experience, the complexities of asymmetrical combat, and military gender norms all combine to create a "perfect storm" for mental health issues. By the way she frames the argument, it is not hard to conclude that all who serve return broken or potentially so.

Holyfield opens *Veterans' Journey Home* by exploring why her interviewees joined the military, what they experienced, and how those experiences reflect a new military at war. She then focuses on the "hazing rituals and degradation rituals" imposed on her subjects during their transition from civilian to soldier in basic training, which she uses to show how the military inherently affects the soldiers the reader met in the previous chapter. She then delves into the professional language of her field, which allows her to argue that the uncertainty and intensity of asymmetrical warfare are contributing to the mass diagnoses of the wars' "signature injuries," posttraumatic stress disorder and traumatic brain injury. She also dedicates two chapters to exploring how gender norms within the military structure also work to set the conditions for the epidemic of traumatic injury.

Although this argument may be objectionable to some, it is at least coherent, which cannot be said for the second half of the book. In it, Holyfield sheds her objective, critical, and scholarly tone, pursuing instead a subjective and personal narrative to detail her experience at various "Vets Journey Home" psychological retreats. This approach allows her to embrace the victim narrative while detailing the history of the program and the rituals and techniques meant to manage it without having to examine the evidence or science behind the approach. The final chapters are merely pages of long quotations from her interviewees and her students in response to a question on the importance of veterans sharing their stories, and in response to the attacks on 9/11, respectively.

Trish Wood takes an entirely different approach than Minear and Holyfield in *What Was Asked of Us*, and the result is equally as different. Unlike them, aside from an overall introduction and brief chapter introductions, Wood does not interpret the interviews in her work at all. Instead, she provides what appears to be an unvarnished compilation of the actual words as they were stated by those who said them, allowing the reader to interpret for themselves. "These veterans do not share a single view of the war they fought," she writes. "They have different opinions about the wisdom of the initial invasion," she adds, noting that a very small number stated frustration with the failure to locate weapons of mass destruction, and a higher number who were proud of the good they were able to do for the Iraqi people.

More important than the background and perspectives of the participants is how Wood presented the trauma they may have experienced and how that affected them. She states that "most veterans in this book will say they were changed in some way by their war experience." But at the same time, she seems to address the hero and victim narratives directly by stating that only some are "struggling with post-traumatic stress disorder," and that none "asks for recognition" beyond being understood.

Eschewing these narratives as gross generalizations, readers can easily find evidence of an emerging notion of personal growth not unlike that described by General Mattis. Personal growth as a theme is not as convenient as the broad and comfortable hero and victim narratives because it does not necessarily align with political worldviews, it does not provide easy shorthand, and it requires deeper investment by the listener. For these same reasons, identifying the theme of personal growth as one of many possible responses to combat experience allows those who served to understand their experiences on their own terms and not on the terms of the broader civilian society.

The emerging theme of personal growth from difficult and traumatic experiences begins with the idea that many in Wood's book saw their wartime experience not as something that harmed them or made them better than their peers but rather as an edifying experience in their lives. Sometimes that edification came in the form of simply serving something greater than themselves, and even being willing to die in that service, and sometimes it is a rite of passage that is facilitated by hardships encountered in combat and its aftermath. Similarly, they do not dwell on being victims of their experiences or being entitled to services because they volunteered.

Like the other authors, Wood includes a section on why the participants joined the military. Adrian Jones and Dominick King joined the marines because family members had also served. King admits to romanticizing combat as a "time where the best of you can shine forth," even though his actual experience was more difficult. For Toby Winn and his twin brother, joining the Marine Corps was the only thing they ever wanted to do because marines were tasked with the most difficult missions.

Maria Kimble and Jason Smithers joined to change the trajectory of their upbringing. "My father was an alcoholic," Kimble says, adding, "and I think other people like me joined the military to escape and start a new life." Similarly, Smithers explains, "I joined the Marines to better myself and prove to everybody that I wouldn't be in trouble like the rest of my family, like going to jail, things like that."

Jeff Englehart joined the army after 9/11 because he was in a "dead-end job" and was "desperate to get out of the United States and travel." While he admits that he was "trying to get [his] life in order," he rejects the idea that he joined "to be a man." Similarly, Garett Reppenhagen describes difficulties stemming from his father's military regimentation and discipline in his upbringing, as well as some things that he brought on himself: "I was twenty years old, I had three jobs, I was a high school dropout, I had a child from a girlfriend that I wasn't married to and I was paying child support, I owned a house and a car but I had to file for bankruptcy, and I was in a dead-end town and a dead-end situation." Eventually, he realized that "standing around in bars talking shit about the government wasn't getting me anywhere."

While other authors and editors have focused on using the oral history format to delve into the politics of service members who are against the war or who were conscientious objectors, including Larry Minear, Wood's transcriptions show very little focus on the politics of the invasion of Iraq or the broader war. There are statements for and against the war, but most of those interviewed focused on their experiences in combat and what they meant to them individually. As young men and women who had survived things that most of their peers had not, they tended to speak about these experiences in ways that many their age could not, they were transformative, and in a culture that values individuality so greatly, they were unique.

One way that these ideas of combat experiences facilitating personal transition and the rarity of those experiences when compared to the broader civilian society come through was in discussions of the loss of innocence. Although a loss of innocence often broadly aligns with the victim narrative, many interviewed in Wood's book discuss the concept with more complexity and nuance. For some, the loss of innocence they experienced was the catalyst for deep and substantial personal growth, and at the same time it was normal, not anything that made them either a hero or a victim.

Jason Smithers states his position clearly: "I don't think I lost anything. I think I gained something." Showing one of the real values of the oral history format, Paul Rodriguez seems to be working through his feelings on the pages of Wood's book: "I'm not going to say I lost my compassion. But . . . I've lost my . . . you know, Adam and Eve lost their innocence when they ate the apple; there was no turning back." Dominick King admits that the war had an effect on him. "I am changed . . . It's not that, that I've changed in a negative way," he says, "I just can't relate with the average college kid anymore." And Daniel Cotnoir says, "no one wants to lose their innocence, but I wouldn't take mine back."

The loss of innocence theme was also discussed in a less explicit way, namely, through participants discussing their experiences in terms of a rite of passage into adulthood. Garett Reppenhagen addresses the issue directly when he states that some of the people he served with "were excited about [being in combat for the first time] because they felt like they went through some rite of passage." Although Reppenhagen rejects the idea that he was experiencing the same personally, others were more open to the idea. "The youngest guys, the single guys, absolutely wanted to go into Iraq," Michael Soprano says in his interview. The difficult and traumatic experiences that others guarantee will result in insurmountable stress disorder were actually the ones in which Adrian Cavazos found hard-earned growth. Referring to a time when one of his fellow soldiers was killed, he says, "at that moment it changed everything for us, and we became hard because we were still young and innocent and just new cherries, if you will, and after that had happened we had become different soldiers."

But even though they endured violent and traumatic experiences, they do not define themselves as heroes. "It's funny who ends up being awarded the medals in wartime," Justin LeHew says. "It isn't the guy you expect it is going to be," which he describes as "the best leader," "the guy that everybody thinks is gung ho," or the one who "everybody would say is the guy who would win the Medal of Honor." In his experience, the hero turned out to be "the kid who basically scored just enough to get into the military," but who stepped up when he had to.

Adrian Cavazos gets to the heart of loss of innocence and heroism when he describes his reaction to the deaths of his two friends, Rincon and Creighton. Referring to Rincon, he says, "he was young and he didn't know a lot about parties and about drinking and he was very innocent," so they treated him "like a little brother." Creighton had been his best friend and they had known each other before the military. "We were young," he says, but in death he remembers his friends not as innocent or as youngsters, but as men who should be remembered for their sacrifice. "Those men—they died," he says. "They died fighting for us. Isn't that better than dying in a car accident or falling in a plane crash or getting hit by a drunk driver?" Relating their deaths to himself, he says, "if I'm going to die, let me do it serving my country or let me do it doing something heroic, something where they can say, he died doing something for someone else, not something selfish or not in some freak accident."

In fact, many participants in Wood's book describe a similar comfort with the possibility of their own death, though they do so in almost explicitly mundane terms. Jason Neely says that he had "pretty much resigned [himself] to dying," and that he "was cool with it." Michael Soprano, Neely's friend, has memories of a corresponding realization: "I remember talking to my friend Neely and saying, 'You know I just know that I'm going to die. I just know it.'" Joseph Hatcher explains that he preferred to be the one to take the risk instead of others in his unit. "I loved the guys—well, most of the guys—I served with . . . I was glad it was me on the back instead of them. It was the worst Humvee we had and, to my knowledge, probably one of the worst Humvees there, I'm glad it was me and not them." Referring to participating in funerals for young marines away from the Iraqi theater, Joseph Darling explains it more globally, explaining that laying his fellow marines to rest stateside made him feel like he should be back in Iraq serving with them. "If I died, I died," he says. "That's my job. That's a big part of why I joined."

Just as they discussed their willingness to die in the service of something bigger than themselves without seeing themselves as heroes, they also never described themselves as victims. While some admitted to having some difficulties processing their experiences, the few who addressed the stress of their traumatic experiences tended to reject that they had been affected at all. Michael Soprano says, "It didn't really affect me that much," about a horrific scene he had just described, while others were more explicit in their thoughts and feelings about the things that others would assume to be traumatic.

Mario Mihaucich says that he actually thought there was something wrong with him because he did not have any "bad thoughts" about his combat experience. "Maybe I really was a cold, heartless son of a bitch or something," he says. "I don't have nightmares. I have memories, but I can talk about them." In fact, beyond being a negative memory, he sees some real positives in his experiences. "I thought it was the greatest thing that ever happened to me, and I think about all the people that live day to day and never do anything significant," he says. "I did something significant, and I feel good about that," no matter what people think the politics of the war will be in the future. Even so, he does recognize that it is complex. "What war isn't psychologically taxing?" he asks. "I think some people just don't know how to process it."

Joseph Hatcher, who was one of the few interviewed who openly opposed the war in Iraq, asks rhetorically, "am I allowed to like it? Am I allowed to enjoy that endorphin rush?" Even though he was against the war, he does not feel guilty, but justified, for his actions there. "The people I shot at knew the risks they were taking," he says, but primarily he admits that he has not really been able to process his experiences yet.

Others tend to agree in their interviews, but even in doing so they betray a degree of conflict in how they view their experiences. "I haven't struggled to cope with society or anything," Matthew Winn, Toby Winn's twin brother says. "I'm drinking. I'm not drinking too much . . . I still jump every now and then if I hear a loud bang or something . . . But I think it's getting better." Dominick King says that he usually does not tell others about what he did in Iraq because he does not want them to interpret his experience through their perspectives. "I was picking up dead bodies," he says. "I don't want to think of myself as a victim. I want to think of myself as somebody who's actually privileged to have a role in something that's changed the lives of so many people." But it is still complex for him. "When I first got back, I felt lucky to . . . to have a story that no one else does," he says. "But then there was also the resentment for me having to bear this whole burden for everyone else back home who, you know, just wants to go to school and get drunk and party."

Through quotes like these, those interviewed in *What Was Asked of Us* show that their combat experiences do not reflect easy narratives of heroism and victimhood that are often reflected in country music, popular culture, and political commentary. They are complex, personal, evolving, and cannot be forced into broad political and cultural narratives, including a narrative of personal growth. Perhaps none interviewed expresses this complexity better than Travis Williams when discussing the death of his brothers, his survival, and the choice he was faced with: "I can drink myself into the ground and turn into fucking nothing or I can take this experience and build off of it and tell the story. I can honor their memory through living my life better because of that, instead of going the route of 'somebody owes me something.'" In fact, he did feel this entitlement initially. "I thought people should fucking feel sorry for me and if I want to drink, then that's what I'm going to do, but I realize that nobody owes me anything and it was my choice to go to Iraq and what happened, happened."

7

On Chickenshit

The Master Narrative

HOWEVER much society, culture, and warfare change, there seems to be a constant narrative that defines a military in combat: chickenshit, and soldiers of every generation can give examples of it even if they have never heard the word. It is the superfluous tasks that senior officers require of their underlings during wartime boredom, the trivialities of rank structures foisted on them in the name of military discipline, the infuriating injustices that seem to occur so easily when the two are matched with personal insecurities.

Paul Fussell, in his book *Wartime: Understanding and Behavior in the Second World War*, offers a thorough study of chickenshit in a chapter called "Chickenshit: An Anatomy." In general, Fussell describes chickenshit as a term that does not merely "imply complaint about the inevitable inconveniences of military life: overcrowding and lack of privacy, tedious institutional cookery, deprivation of personality, general boredom." Rather, chickenshit describes the behaviors that make "military life worse than it need be." Specifically, chickenshit is:

> petty harassment of the weak by the strong; open scrimmage for power and authority and prestige; sadism thinly disguised as necessary discipline; a constant "paying off of old scores"; and insistence on the letter rather than the spirit of ordinances. Chickenshit is so called—instead of

> horse- or bull- or elephant shit—because it is small-minded and ignoble and takes the trivial seriously.

Chickenshit is easily recognized, Fussell writes, "because it never has anything to do with winning the war."

Fussell's phrase offers a timeless lens through which subordinates understand the actions and inaction of their seniors in military hierarchies, on that would ring true for any generation of military members, though in different ways. Soldiers who have fought in Iraq and Afghanistan have a remarkably different worldview and experiences than those in Fussell's book who served in World War II. The type of warfare, the American culture they joined the military from, the way they consume media, and the political climate have all changed between the last truly modern war and the first truly postmodern war.

Accordingly, Fussell's depiction of chickenshit in World War II requires an update, one that shows how the all-volunteer force, the rise of social media, the increased divide between civilian and military cultures, and the professionalization of the military have changed the experience of the soldier on the battlefield tremendously, including the ways that they experienced and responded to chickenshit. In its most recent manifestation during the wars in Iraq and Afghanistan, and immediately after, chickenshit serves as a master narrative, sometimes unifying and sometimes serving as a reference point for the various other cultural and political narratives about the wars.

While Fussell refers to chickenshit as an aspect of military culture that perpetrates power of the strong on the weak through dangerous or mundane assignments, it manifests somewhat differently in the literature of the wars in Iraq and Afghanistan. While the authors certainly use chickenshit, and more often bullshit, to describe the mundane emotions and trivialities of wartime, and also the exercise of power by a senior over a subordinate, in these works the phrases carry more weight. In these wars, instances of chickenshit evoke a sense of dishonesty often used in political rhetoric to persuade, and political cowardice that allowed for a poorly planned invasion and the belief in a painless war, all resulting in a risk-averse approach to combat. And, because those who served in the military during these wars did so voluntarily, it helps explain why those who endure chickenshit in the military can complain about, but how they do not see themselves as heroes or victims for doing so.

Fussell's most interesting examples of chickenshit come from fiction authors who had written about their war. These authors were "precisely the types that are chickenshit's eternal targets, bright Jewish boys like Norman

Mailer and Joseph Heller, or intelligent sarcastic kids from good colleges, like Kingsley Amis." Through their fictional accounts, these writers were the spokespeople of a disgruntled group of conscripts who were unwilling or unable to register their complaints for themselves. Timothy Corsallis's poem, "What I Never Saw," for instance, is a brief description of the lack of high purpose he expected to encounter. Amis catalogs the chickenshit of Major Raleigh in *My Enemy's Enemy* so thoroughly that it would be humorous if the stories of bureaucracy and retribution did not represent global truths of wartime experience. And, of course, Joseph Heller's *Catch-22* is a case study in chickenshit.

To examine such novels for today's wars proved to be difficult, though, well into the war effort. Former army officer Matt Gallagher wrote in a 2011 piece for the *Atlantic* that while literary markets are saturated with war memoirs and other nonfiction works about war, "even the most avid bookworm would be hard-pressed to identify a war novel that could be considered definitive of the new generation's battles." Perhaps a war needed to be over before it could be examined, he suggests, or perhaps there was just no market for fiction of his war in Iraq. Gallagher settles on what seems to be the beginning of an explanation with a terse comment from a colleague posted on Twitter: "Our wars lack collective national will, [there's] confusion over purpose, and are protracted."

The lack of literary fiction associated with contemporary wars was not a new complaint when Gallagher wrote his piece. Tom Wolfe wrote of similar concerns in his classic *Harper's* essay, "Stalking the Billion-Footed Beast," which was mainly dedicated to chastising the writing establishment for getting away from realism in fiction. He writes that publishers

> had their noses pressed against their thermopane glass walls scanning the billion-footed city for the approach of the young novelists who, surely, would bring them the big novels of the racial clashes, the hippie movement, the New Left, the Wall Street boom, the sexual revolution, the war in Vietnam. But such creatures, it seemed, no longer existed.

Wolfe famously proposed New Journalism, an approach to nonfiction that was bolstered by the principles of literary fiction, as the solution to the editors' quandaries.

A similar mingling of genres has occurred since Wolfe wrote that, and it could be argued that his approach has much to do with what is happening in the wars in Iraq and Afghanistan. These wars have been prosecuted during the information age, with constant updates fed to a hungry global public through embedded reporters, with soldiers filling the remaining void with blogs, pictures, and videos. Wolfe's New Journalism encouraged more access to the authors' opinions, their senses, and even approves of a novelistic approach to nonfiction, and the new media of wartime bloggers in these wars is the logical extension of Wolfe's approach.

The most famous of these blogs have become books, allowing them to exemplify changes in media as well as changes to the wartime memoir. In the best examples, these books are a coherent volume based on the blog posts. At their worst, they are merely edited and bound printouts of the blog posts with no real context or depth. Although there are numerous examples, Jason Christopher Hartley's *Just Another Soldier: A Year on the Ground in Iraq* and Colby Buzzell's *My War: Killing Time in Iraq*, Benjamin Tupper's *Greetings from Afghanistan, Send More Ammo: Dispatches from Taliban Country* and Matt Gallagher's *Kaboom: Embracing the Suck in a Savage Little War* effectively represent early examples of the blog-to-book format.

These books perform many of the functions that fictionalized accounts served during Fussell's war. Of these works, it is Hartley's strength as a writer that sets him apart from his peers. He thrives in the central conflict of his work, that he is a voluntary infantry soldier deployed with his unit to Iraq while rebelling against the constraints and structures of the military. The only time that he does not seem to be involved in this conflict is when he is focused on an external conflict, that of life and death.

Buzzell's book is at times funny, irreverent, cynical, and deeply representative of his peers on the battlefield. However, if the title of his work, *My War: Killing Time in Iraq*, also indicates some internal struggle while in training or deployed, his book is almost profoundly devoid of such examinations. If Hartley's book is about defining conflicts, developing them, and eventually mastering them, then Buzzell's book merely describes the conflicts he sees, not seeking to understand them, or even recognizing the conflicts he creates in telling his own story.

Benjamin Tupper's *Greetings from Afghanistan, Send More Ammo* is a very straightforward representation. He covers his experiences as an Embedded Training Team leader in Afghanistan from 2006 to 2007, detailing the intensity of combat and the boredom of daily life. Conversely, Gallagher's *Kaboom* successfully combines the finest innovations of blogging with the finest aspects of traditional memoir writing, while avoiding the traps of each. Of the four books, Gallagher's is authentic, fully and successfully updating for his generation what Amis, Heller, Evelyn Waugh, and others did for their generation.

Collectively, these books show that chickenshit is still a constant burden for those who serve in the military, and they are even peppered with similar barnyard language. Although they sometimes use the word *chickenshit*, they do not do so as Fussell did to explain the ridiculous ways that the military power structure exhibited itself. Instead, they use it and its many variations—horseshit and bullshit, mainly—more generically to describe different forms of fear and pettiness.

Buzzell provides several important examples of these evolved definitions of chickenshit. About his unwillingness to deal with an alligator, Buzzell writes that his fellow soldiers told him to "stop acting like a chickenshit." And later, he writes that attacks on his platoon had mostly been the "chickenshit hit and run bullshit" of an insurgency, noting, "every time we'd get hit, they'd be nowhere in sight." Hartley adds to the conversation by providing similar examples. He wants to make it clear that he is not whining, explaining that "bitching is the only truly inalienable right of the soldier," that even though he loves the military, "that doesn't mean I have to enjoy the taste of horseshit." Later, he channels Fussell when describing being busted for a curfew violation during training. "I'm a thirty-year-old infantryman who will soon be given the ultimate authority of killing anyone I adjudicate to be a threat," he writes, "but until then I have the same curfew that I did when I was in junior high." His punishment for missing his curfew causes him to use a common variation on chickenshit, noting, "plane tickets and hotel rooms have already been paid for by my father and sister so they can come up for the shitbird deployment ceremony."

Beyond these basic examples, though, those who served in the wars in Iraq and Afghanistan do not dwell as long or as deeply on the unbearable chickenshit as their predecessors did in fiction. To understand how chickenshit in the military evolved, it must be examined more broadly. A crucial text for this is *On Bullshit*, by Harry Frankfurt, a moral philosopher who uses the tools of linguistic analysis to plumb the depths and find the true essence of his subject, deeper even than Fussell delves regarding chickenshit. In *On Bullshit*, Frankfurt offers a definition that is strikingly similar to Fussell's, referring to the British military usage of "bull." He writes:

> according to the [*Oxford English Dictionary*] . . . refers to "unnecessary routine tasks or ceremonial"; excessive discipline or "spit-and-polish";

> = red-tape. . . . Here the term *bull* evidently pertains to tasks that are pointless in that they have nothing much to do with the primary intent of justifying purpose of the enterprise which requires them. Spit and polish and red tape do not genuinely contribute, it is presumed, to the "real" purpose of military personnel.

Frankfurt has a linguistic advantage over Fussell, though, in that *bullshit* can also be used as a verb, while Fussell's chickenshit is only a noun, allowing Frankfurt to better study bullshit from a prescriptive perspective. In *On Bullshit,* Frankfurt asserts that the act of bullshitting someone has some of the characteristics of lying to them, but at the same time it is somewhat "short of lying." To explain this concept, Frankfurt asks his reader to imagine that a person tells them a lie about how much money they have. In doing so, they are lying twice, once about the amount of money, and once by indicating that they believe what they are telling you is true. A bullshitter, Frankfurt argues, is guilty only of the second deception. Frankfurt later gets to the heart of his definition: it is the "lack of connection to a concern with truth—this indifference to how things really are—that I regard as of the essence of bullshit."

In a central summation of Frankfurt's theoretical approach to the meaning of bullshit, he turns toward cultural application. "The realms of advertising and of public relations, and the nowadays closely related realms of politics, are replete with instances of bullshit so unmitigated that they can serve among the most indisputable and classic paradigms of the concept," he explains. But even through the virtue of being a bullshitter, and therefore not as bad as a liar, it is still true that the messenger aims to "get away with something." It is also possibly true that "although it is produced without concern for the truth, it need not be false." In many ways, this description fits the way war in Iraq is often described as being sold: "[the bullshitter] does not care whether the things he says describe reality correctly," rather, "he just picks them out, or makes them up, to suit his purpose."

The debate over the selling of the war in Iraq has become an enormous part of the narratives that make up America's political landscape, but it also reveals another element required for updating chickenshit. In the context of these wars, chickenshit is not shown to only apply to situations where military leaders exact revenge or assert authority over their subordinates. It also refers to a similar cowardice from the nation's civilian leadership during wartime.

In 2004, a University of Georgia sophomore, Bradley Alexander, took to the pages of a conservative magazine to excoriate his professor, John Morrow, for his introductory lecture to a history class. Alexander was uncomfortable with Morrow's views on the military conflicts in Iraq and Afghanistan, how they applied to the greater topic of the nation's history, and particularly Morrow's use of foul language, including his use of the word chickenshit. In response, Morrow stood by his critique, including his usage of chickenshit:

> for students of military history, it is one of the milder epithets, and if they are going to take a course on war, they are going to encounter a good deal of profanity. Veteran Paul Fussell's excellent work *Wartime* defines "chicken shit" in reference to the bureaucratic garbage that soldiers of World War II encountered.

But Morrow expands the definition of chickenshit and connects it even further to the moral application Frankfurt provides through his examination of honesty in bullshit by stating that the term "also refers to people who do not have the courage of their convictions," and that he "applied the term specifically to President Bush and Vice President Cheney, who avoided service in Vietnam." Morrow found it ironic that Bush avoided service by joining the National Guard, but "now had no difficulty sending the Guard to extended duty in Iraq." He continues, writing, "I further condemned this administration for ignoring and even dismissing its best military brains, who informed Rumsfeld, Cheney, and Bush that we needed 'more boots on the ground' in Iraq, a clear set of goals, and an 'exit strategy' before we entered the war."

By citing this lack of courage of conviction, Morrow connects chickenshit to a its relative, "chicken-hawk." Fairly or unfairly, the term has been effectively leveled against Bush and Cheney because of the perception that President Bush used his family name and connections to avoid active military service, and the fact that Dick Cheney pursued multiple student deferments instead of participating in the draft during Vietnam. But even if the president and his lieutenant had served in wartime, the chicken-hawk critique could still apply more broadly to the civilian administration, including all branches of government. The administration had the support of nearly every major political leader in Congress, and from both sides of the aisle, and all sides seemed determined to prosecute the war, to make it as painless as they possibly could, and to spare no expense in doing so. As Morrow points out, there proved to be no mission, no tangible goals, no timetables, and no exit strategy. Lawmakers on both sides chose the chickenshit route of voting for a war without declaring war, and sending soldiers, sailors, airmen, and marines into combat without an idea of what victory would look like or what it would cost, and failing to mobilize the American people behind them.

This chickenshit, much like the chickenshit Colby Buzzell described earlier, references fear. But in this case, the fear was not that of soldiers on the front lines, those who could be excused for being afraid. Rather, in this usage chickenshit describes those who were afraid of political minefields, of presenting a timeline, goals, and an exit strategy, afraid of being painted as being "against the troops," and afraid of taking a nuanced position on the war, for or against. Here it describes leaders from both parties who took the chickenshit way out of a very difficult problem facing the nation, while seeking opportunities to blame others for their decisions or inaction. Leaders from the political left could claim that President Bush lied to send a nation to war, while leaders from the political right could claim that their opponents failed to support the war effort. This form of Frankfurt's bullshit descended from the highest levels of civilian leadership all the way to those on the battlefield, as these contemporary memoirs point out, even though they are more likely to refer to this form of chickenshit as "bullshit."

The chickenshit or bullshit as depicted in these books comes in at least two different categories. Often, in the military, bullshit is conducted or enforced not because of unit discipline, but because of the intent of putting forth a desired image, the things that Frankfurt describes as "unnecessary routine tasks or ceremonial" or "spit-and-polish." But they fit the rest of his definition as well, in that they are pursuing a certain perception even when the reality is likely the opposite. This sort of deception is an age-old military custom, one that is no less defined by Fussell's chickenshit as it is defined by the "short of lying" in Frankfurt's bullshit.

The second most common form of chickenshit listed in these books is far more related to that described by Morrow, the moral cowardice that comes in the form of a deeply risk-averse chain of command to the lowest levels of combat forces. Since America and its policy makers were morally and personally disconnected from these wars abroad, there was an unprecedented amount of fear associated with the hardship of combat for those they send. It started at the highest levels, with more expendable contractors taking the role of soldiers on the battlefield, and it went all the way down to the lowest ranks, with officers and noncommissioned officers making decisions that infantilized soldiers on the front lines.

Regarding the spit-and-polish pretensions described interchangeably as "bullshit" or "chickenshit" by Frankfurt and Fussell, these memoirs provide countless examples. Buzzell complains, "even though we're deployed in a 'combat zone' we still do a lot of the same bullshit that we do in the rear," referring to the type of things they might do at their home station. Gallagher quotes one of his soldiers complaining about being back on their Forward Operating Base, or FOB, around all of the brass. "I hate it here, sir," his soldier says to him. "We always get bitched at for garrison bullshit here," he adds. Hartley provides a poignant example that might be believable only to those who have been in the military, referring to a memorial ceremony for a fallen soldier named Akintade, In Hartley's description of the scene, Akintade's gear is on display in front of the unit while the sergeant major, inspecting it, finds an issue. "Everyone in the formation had the black plate for their night-vision goggles on their helmets, but the helmet on display to represent Akintade didn't." This problem was easily explainable, the plate was a sensitive item that needed be accounted for even after a soldier has been killed, but that did not alleviate the sergeant major's concerns for the lack of uniformity.

Other examples of chickenshit described in these books fall squarely under the banner of the type of risk aversion that infantilizes a combat force. Tupper writes, "absent any real threat, the powers that be at Bagram have dug up some B-list threats, and have established some ridiculous safety polices with the intent to keep us all on our toes." He lists the enforced use of seat belts on the base, as well as reflective belts and three-mile-per-hour speed limits on post.

In describing a scene where he was encouraged to line his truck with sandbags to defend against a mine attack, Hartley invokes the barnyard term to describe the questionable defensive measure, noting, "all of the guys in my truck agreed that we would forgo the sandbags-on-the-floor bullshit and just live dangerously." Capturing the idea of chickenshit as risk aversion perfectly, Gallagher cites an inscription on the wall of a portable toilet: "Wash your hands before returning to war! Thanks.—Command Sergeant Major." While they were deployed to combat zones, these writers point out, risk aversion has become the watchword from the highest level of policy makers all the way down to the front-line enforcers.

Chickenshit, as it is described by these authors, was at its worst when it was a combination of both of these forms, the image-driven superficiality and the inherent risk aversion. Benjamin Tupper writes a particularly poignant piece on the passing of an admired noncommissioned officer, Sergeant First Class Deghand, who was killed during their deployment. On the surface, Deghand had been moved due to interpersonal squabbles, but Tupper describes the move as being based on the fact that he was a "straight shooter [who] didn't put up with [higher's] bullshit." One can never know if Tupper's description is true, but the settling of old scores by the powerful is exactly the kind of thing Fussell would describe as chickenshit.

Gallagher's best examples of the combination of the two types of chickenshit describe his view of the generational gap between the Cold War–era leaders and their counterinsurgency operators on the ground. Those who were visiting this chickenshit on the lower ranks are field grade officers, some of whom were great and excellent at their jobs, but then there were the others who seemed "intent on riding the bureaucratic beast in all its protectionist glory." These were the ones who were "in too-clean uniforms, criticized trivial things, like soldiers not shaving enough or not wearing a full uniform while they walked to the shower, and then drove back to the [base] in time for dinner."

The new capacity of deployed service members to write about their experiences immediately through blogs and other options came with another obvious set of tensions within the military command structure, opening a new possibility for chickenshit that could happen only to this generation of soldier. Combat units traditionally focus on discipline, particularly with respect to protecting operational security through avoiding public disclosures of tactics or locations. But this discipline can also be used to limit criticism of the chain of command. The development of this tension between the good of the collective and the rights of the individual provided another interesting possibility for wartime chickenshit, particularly with respect to how their blogs were seen by their higher-ups.

Tupper addresses the subject only at the end of his book, suggesting that some were uncomfortable with his writing his blog, but no one tried to stop him. Unlike Tupper, Gallagher had long observed the military's published policy on blogging, which mandated that commanders clear any of their subordinates' posts before publication. But this changed for him when his battalion commander directly threatened his career, apparently because he had decided to leave the military after his contractual commitment. In what Gallagher describes as a mixture of rage and exhaustion, he posts a blog questioning his command's competence, their ability to reason, and their concern for things that truly matter. Accordingly, his commander promised him that he would receive "the next bullshit tasking" that came down, removing him as platoon leader. Gallagher described this as "professional bullying," but Fussell might describe it as chickenshit.

For Hartley, the consequences for getting caught blogging could hardly have been worse. Although he chafed under the authority granted to his leaders over him, he genuinely appreciated the military and intended to excel at his profession. After being reprimanded for breaking curfew before the deployment, Hartley makes what he sees as the mature decision to discontinue his public blog, only to keep it alive by setting it to private. Almost a year after he had stopped blogging publicly, and closer to the end of his time in Iraq, Hartley drunkenly decided to take his blog live again, complete with all the entries that he had not published to date, and then almost immediately going on leave from Iraq. For this, he was shuffled through a disciplinary process full of the worst kinds of chickenshit, which he describes in full detail.

Buzzell's story of tension with his command about his blog exhibited in *My War* shines a less sympathetic light on the author. As he describes it, he was given a great deal of latitude with respect to his writing. His commander was even a fan of Buzzell's prose, inviting him to his office to discuss the glories of Hunter S. Thompson with him and complimenting his writing. When concerns for operational security were pointed out, Buzzell even notes that his commander was right and agrees to scrub information the enemy might find useful.

A few weeks later, though, Buzzell, was called into his sergeant major's office and was given a similar speech. The army was not interested in infringing on his freedom of speech, he was told, but they were still trying to enforce operational security. Because Buzzell's blog was starting to get more international attention, he was informed that he would no longer be allowed to go on missions off the operating base. The punishment seemed draconian and was eventually reversed after media pressure, only to be replaced by an approval process by his chain of command for each of his future entries.

This punishment was hardly as harsh as the retribution delivered in response to Gallagher's and Hartley's blogs, but Buzzell still bristled under the restrictions and decided to discontinue his blog. While on leave from Iraq, he reached out to Jello Biafra, a punk musician who was famously against the war in Iraq, and asked him to write a message to post on his blog. He had not cleared it with his command, but it would not have passed muster anyway. In a way that perhaps only Joseph Heller could fully appreciate, the soldier who was treated fairest by his commanders visited chickenshit on them in return.

In fact, the chickenshit that all of these authors describe was tempered with something else that would be terribly uncommon in the fiction from World War II. During previous wars, the nation had instituted a draft, meaning that a much larger percentage of the national population was affected by the war effort, and these books include many statements regarding that civilian-military divide. Gallagher writes, "in a volunteer military, we fought for the nation, not with it." Or when he copies a comment from a portable bathroom wall: "America isn't at war. Soldiers are at war. America is at the mall."

Gallagher attributes this divide to the national leadership at the time of the Iraq War. "I slept through 9/11," he writes. "Nine days later, I yawned along with most of my peers as the president asked for our continued participation and confidence in the American economy." While the military was preparing for war, the president wanted the nation "to keep shopping."

But even as they complain, they wear it as a badge of honor as well. Hartley points out that in an all-volunteer force, some of the best stories are those that could be described as chickenshit: "when soldiers tell stories, the ones they are most proud of are those that involve things sucking." Even the subtitle of Matt Gallagher's memoir, *Embracing the Suck in a Savage Little War*, shows this dynamic. Perhaps the best explanation of this comes from Hartley when he writes, "being a soldier is to live in a world of shit," he writes. "You're constantly surrounded by assholes; you have to endure an unending amount of bullshit from you leadership, military regulations and paperwork, and stupid training missions."

Hartley's description, which he labels the Tao of Soldiering, is the epitome of chickenshit for soldiers in the wars in Iraq and Afghanistan. It is the same old nonsense that has been endured for generations. It is different for those serving in places like Iraq and Afghanistan because the culture has changed, because the ways they communicate and express themselves have changed, and because risk aversion has permeated policy from the highest levels down. Perhaps the biggest difference of all, though, is how this generation responds to it. To them it is just as onerous and infuriating as it has been for every other generation, but because they are a generation that volunteered for their service, enduring it does not make them heroes and it does not make them victims.

8

In the Mushy Influences of the Current Times

New Narratives

"THE REAL WAR will never get in the books," wrote Walt Whitman of the Civil War. "In the mushy influences of the current times, too," he wrote, "the fervid atmosphere and typical events of those years are in danger of being totally forgotten." It seems that not only was the country moving beyond the prolonged national struggle emotionally, but the truth would also be veiled by politics, memory, or other motives. Whatever the reason, he knew that the darkest tragedies experienced by those of his war would be lost by the literature that would follow, unexplainable to those who never served or saw it firsthand.

In *Wartime*, Paul Fussell uses Whitman's quote to expand the same assertion for his war, and in more detail. Embracing Whitman's lament, Fussell argues that no book could truly capture the experiences of those who served in combat in World War II for those who did not serve—not journalism, not memoirs, and certainly not novels. The experiential divide between civilians on the home front and soldiers on the front lines was so great, and the cultural bridges that should have connected them so shoddy, that the real story of his war simply could not be reliably conveyed through literature of any kind.

Fussell offers a number of arguments to back up his assertion. He contends that his war was being "systematically sanitized" in such ways that prevented the public from ever being able to understand it, and those who were doing the fighting knew it. This happened through media portrayals of the war back home that were inadequate at best and corrupted at worst, with Fussell arguing most often for the latter. Referring to the evidence in pictures and stories that could have shown the devastating cost of war, Fussell asks, "how is it that these data are commonplaces only to the small number who had some direct experiences of them?"

This sanitization was due in part to "the normal human talent for looking on the bright side," and in part to the practice of members of the media who experienced the horrors of war alongside service members but "kept quiet about them on behalf of the War Effort." Fussell criticizes John Steinbeck's dispatches for ensuring that "the morale of the home front must not be jeopardized by an eye-witness saying so." He chastises Ernie Pyle for following the rules required for the access he was given, meaning he was to "reveal only about one-third of the actuality and just like the other journalists fuel all the misconceptions—that officers were admired, if not beloved; that soldiers, if frightened, were dutiful; and that everyone on the Allied side was sort of nice." By doing so, they "could give the audience at home the impression that there were no cowards in the service, no thieves and rapists and looters, no cruel or stupid commanders."

For those who served in the war and understood the way it was being transmitted home, this divide between experience and perception "augmented their sardonic, contemptuous attitude toward those who viewed them from afar." Such sardonic and contemptuous tones can be seen in a handful of novels to come out of World War II, mostly from authors like Joseph Heller, Kurt Vonnegut, and Evelyn Waugh, but even so, Fussell insists that "the real war is unlikely to be found in novels." Novels, he writes, "must exhibit, if not plot, at least pace, and their characters tend to assume the cliché forms demanded by Hollywood, even new Hollywood." And "despite undoubted success as engaging narrative, few novels of the war have succeeded in making a motive, almost a character, of a predominant wartime emotion, boredom, or persuading readers that the horrors have not been melodramatized."

For both Whitman and Fussell, and for many others who discuss the divide between a people at home and those deployed in military service abroad, these are basic facts of wartime and its literature. For Whitman, this is probably due to a collective need to keep moving forward, and while that exists in Fussell's description of the phenomenon, he is far more willing to blame a sort of cultural cowardice that lends itself to the collective need for avoiding the blame for passively allowing the atrocities of war on fellow citizens. To both, this leads to a wartime literature that is inauthentic, that embraces the active political narratives of the day, and that does not allow for a deeper understanding of the true costs of war to the individual member of the military and to the society who sends them off.

The cultural and media landscapes of the nation at war in Iraq and Afghanistan are remarkably different from those during World War II, America's last truly modern war, to say nothing of the Civil War. Due to the draft, World War II was a more personal war to all of the American public than the wars in Iraq and Afghanistan, which were fought by an all-volunteer force, ever could be. And barring the time prior to the widespread looting in Iraq in April 2003 and after David Petraeus took command in February 2007, members of the media have largely refrained from becoming cheerleaders for the war effort, as Fussell describes in his war.

That is not to say that the experiential divide between those who served in the Iraq and Afghanistan wars and the civilian populace has grown closer. And it is not to say that while there were probably fewer media cheerleaders during the Iraq and Afghanistan wars, the national understanding and conversation has actually improved. These wars merely reflect differences that have for some time still ended in the same results described by Whitman and Fussell.

These differences that achieve the same results arise from a number of sources. The distance between the public and its military has changed during these wars due to the use of an all-volunteer military, making sure that only the smallest percentage of Americans are affected, or more important, have personal access to trusted perspectives from those who have been there. Further, information about these wars has largely been consumed in ways that serve to confirm the political, social, and ideological worldviews of specific segments of national audiences.

However, and quite different from Fussell's criticism of wartime literature in general and of fiction in particular, the wars in Iraq and Afghanistan have already provided for an extraordinary new access to the real experiences of combat, even for those who have not experienced it firsthand. Those seeking to find the real war, not the war as it is filtered through political, social, and cultural narratives, but as told in an honest, forthright, nuanced, and sincere way, can turn to fiction for a largely unmediated variety of stories. Thanks to the fiction writers of the wars in Iraq and Afghanistan, the real war is finally in the books.

Perhaps perfect for explaining America's first truly postmodern wars, many of the best works of fiction about the Iraq and Afghanistan wars are not presented in ways that are "systematically sanitized," as Fussell described the literature of his war, but depict brutal realities of war. They very often challenge many of the widely accepted narratives about their wars as well. And, by using fiction, they explore the truths of these wars in ways that most works of nonfiction, whether memoirs, journalism, or otherwise, have not.

An important aspect of the literary fiction from the wars in Iraq and Afghanistan is that it has been effectively written by both writers who served in the military in the country and by civilian authors. This alone allows for a broader national discussion of the wars, and it allows for the assumptions and narratives of them to be challenged from more perspectives. Unlike the criticisms leveled by Whitman and Fussell about their wars, these books do in fact provide for a common understanding, and they do so through a discussion that is even richer than they imagined for their wars.

Although there are many authors who have written literary fiction involving the wars in Iraq and Afghanistan, four well represent the military and civilian perspectives. From the civilian perspective, Lea Carpenter wrote *Eleven Days* about a navy SEAL who is killed in a scenario similar to the Osama bin Laden raid in Pakistan. Laura McBride wrote *We Are Called to Rise* about how some brought the war home with them, and how that affected lives in Las Vegas, Nevada. Phil Klay, a former officer in the Marine Corps, wrote a collection of short stories called *Redeployment*, which received the National Book Award for fiction in 2014. And Matt Gallagher, a former army officer, has written a memoir on his time as a cavalry platoon leader during the surge as well as a novel set in Iraq a few years later titled *Youngblood*.

In interviews, these authors discussed the state of the political and media rhetoric surrounding the wars, and they comment on how their works fit into it, at times, and at other times, to challenge it. Although they do not share identical perspectives, they tend to agree that literary fiction has not only expanded the current conversation, but it has also allowed both writers and readers to understand deeper and more complex truths about the wars. More important, the wave of fiction from the wars in Iraq and Afghanistan, which each of these authors has contributed significantly to, has allowed for a bridge between the military and civilian communities for the first time since the wars began.

These authors recognized that much of what was presented to the public about the wars in Iraq and Afghanistan, about those who fought them, and about the families they left to do so, did not reflect their personal experiences or the experiences of those who they talked to about the war. The way the wars were presented by the media, through some memoirs, and through other outlets was more sensational, it was more superficial, and it was more political, reflecting more of what the broader civilian public wanted to hear than the true and complex experiences of those who were deployed on their nation's behalf. They recognized this divide, just as Fussell did during his war, but unlike Fussell, they saw an opportunity to bridge it through literature, and especially fiction.

Even while in college and pursuing a commission in the army after the attacks of September 11, Gallagher was aware of how political the discussion of the Global War on Terror was. "I think at that point I was conscious of what Vietnam meant to America, and how the war debate around Iraq was still influenced by Vietnam," Gallagher says, adding, "and how all of the talking heads screaming at each other on CNN were doing the same thing they had been doing in 1969 in dorm rooms." Later, while serving as a Scout platoon leader with Twenty-Fifth Infantry during the surge in 2007, Gallagher began writing a blog for his friends and family back home, and one of his reasons for doing so was to provide his perspective from the front. "The war was very political then, still is, and I had friends all over the spectrum," Gallagher says, "from die-hard liberals who were very anti war, to conservatives who thought that this was the best thing that has ever happened to the Middle East." His blog, which would later become his memoir, *Kaboom*, provided them all an opportunity to read about his firsthand experiences if they wished.

Phil Klay describes his reasons for writing his short story collection, *Redeployment*, about the war in Iraq as related to "the strangeness of coming back to America after being in a warzone," a strangeness that challenged him further as he attempted to write fiction. "For me, writing fiction," Klay says, "you take whatever ideas you have about the world and you throw them in a story," beginning a process that inherently exposes "how flawed you are," which will encourage you to change the choices or trajectory of the characters in your work. It was not just his attempts to better understand his experiences and those of others that amounted to the strangeness he felt, though, it was also what he recognized as "the difficulty of having a discussion about the war that doesn't turn into a discussion about one political side or the other." His experience in the war was more complex than two political perspectives, and he saw fiction as an opportunity to change the trajectory of the conversations.

In focusing on the events surrounding the death of a navy SEAL, especially as it played out stateside, Carpenter wanted to write a book that would provide access to all readers, no matter their experience or perspective. "I really wanted to write a book that someone who does not know the difference between the Army and the Navy could appreciate," she says. But as she immersed herself in the naval special warfare community, she realized that she also wanted to write a novel that those in that community would see as accurate and authentic. "There are a lot of stereotypes that latch on to the military, in general, and to that community in particular," Carpenter says, "and it's a community that has evolved a lot since the first two SEAL teams were commissioned in Vietnam." This evolution, which continues, has been affected by the "tactical and policy noise" surrounding the navy SEALs, particularly after the successful raid that killed Osama bin Laden, and all while the families and the special operations community deals with a generation of loss at war.

Referring to the length of time it took for writers to start publishing fiction of the wars in Iraq and Afghanistan, Gallagher says, "there was a real question about whether there would be a fiction readership for these wars" due to the wars' "political, polemical nature." But he also sees that time as necessary for allowing authors and the country the time and distance from the war they need to process their own experiences, and also to challenge their beliefs against history as it unfolds. However long it took for the wars to produce literature, Gallagher sees the rise of fiction as positive because the fiction coming out of the wars "defies the narrative that America wants to hear."

In fact, these authors challenge the narratives surrounding the wars, and in important ways. Gallagher and Klay largely do so from the perspective of those who had been on the ground in Iraq, giving them the room they need to write harsh and foreign truths about the wars that are seldom presented elsewhere. As civilian authors writing about war, Carpenter and McBride spend more time examining the effects of the war at home, particularly with respect to two prominent narratives from the Iraq and Afghanistan wartime era.

In *Eleven Days*, Carpenter addresses the nation's commitment to the war, which was reflected in how averse to casualties American civilian and military leadership were, even in the special operations community. "When I was three months into writing this book," Carpenter says, "there were two events that happened in the Naval Special Warfare community that I felt I couldn't not deal with." The first was the raid resulting in the death of Osama bin Laden at his compound in Abbottabad, Pakistan, and the second was a "helicopter crash resulting in the then greatest loss of life in the special operations community." Carpenter would make the Abbottabad raid a central part of her novel, but the major loss of life due to the helicopter crash led her to write what she describes as the most meaningful portion of *Eleven Days.*

"One of the people that I spoke to in the course of my research said something that was very interesting," Carpenter says, referring to a member of the special operations community who she interviewed. "One of the things that he said to me was that there was a phrase often repeated, particularly, in speeches delivered to guys before they went out on their first raid, and that phrase was, 'no losses,' the goal is no losses," she says. This was perplexing to him for a number of reasons, leading him to wonder why they were even conducting a combat mission.

To Carpenter, thinking about this conversation revealed an evolution among policy makers away from an interest in high numbers of enemy body counts, as was the case in Vietnam, to one that was focused instead on low numbers of friendly casualties. There could be any number of explanations for this evolution, from a sincere desire not to lose troops to a cynical desire to manage the national mood by limiting the amount of emotional and material investment America made to the war, but Carpenter did not pursue these leads as a nonfiction writer might. Instead of pursuing the policy discussion, Carpenter chose to examine how the meaning of those losses, for both members of the military and the broader American public, might be remembered in a meaningful war.

"In the climactic chapter of my novel I tried to show the stream of consciousness that Jason is having," Carpenter says, referring to her protagonist. "He's on the helicopter flying into Pakistan, and he's thinking about a lot of things—how did I get into this line of work? What does it mean? Where am I going? What's next?" It is in this context that "he remembers this phrase, 'no losses'":

> "No losses," the young guys say, to a man. "Zero casualties." They say this when asked what their goals are before going out the first time. They say this because this is what they're told is the goal. Sometimes Jason wanted to say to his senior officers, "If the goal is no casualties, why are we going on a *combat* mission?" But he never did. Conflating loss with failure wasn't right, but making them distinct was risky. *One million, three hundred forty-three thousand, eight hundred twelve*: a number almost equal to the populations of Oklahoma City and Austin combined. Higher than Indianapolis plus Long Beach; lower than DC plus Jacksonville;

> almost equal to Detroit, Olathe, Salt Lake, and Aurora combined. If you added up the total populations of Grand Island, Rogers, Union City, and Shreveport—and then added in those of Green Forest and Angwin and Willits and Southern Pines—you would arrive at a different number. Olathe, Aurora, Angwin, Southern Pines, Austin, Willits: these were all cities that sent sons to serve and lost them. "A helicopter is a dangerous vehicle to go to war in. When you step into a helicopter, you're taking a risk," said the judge who'd presided over the Sikorsky case, the one who came to squire Sara after Jason moved back east. No successful mission makes the losses matter less. And yet while preserving the memory of lives lost was critical, an ability to avenge them slipped like sand through a jeep's grill. *After the peak there comes the challenge of how to ride down the other side.*

In this passage, Carpenter tries to provide context for that phrase. "I looked at the populations of certain cities in the U.S. history, and I tried to just play around with that math," she says. "If you add up the populations of city X and city Y and then you subtract out the population of city Z then you get this number that stands in correlation to lives lost in" previous American wars. For the civilian public, Carpenter hoped the mathematic approach would be jarring. For those from the special operations community, she hoped to provide "a eulogy for that day," because the names of the cities referenced are the hometowns of the casualties from the helicopter crash.

The connection between the realities and the fiction of the wars that Carpenter provides allows readers to move beyond basic political conversations about them. She does not denounce a policy from a political perspective, but rather simply points it out. And in doing so, she provides it as a part of a bigger discussion about what policies like these say about a nation at war.

In *We Are Called to Rise*, Laura McBride also address aspects of narratives surrounding the wars, also with respect to the nation's commitment to the efforts. Her story takes place primarily in Las Vegas, where McBride lives and works, allowing her to explore the effects of the wars on the larger civilian community. While she never directly addresses the topic of how fully the nation was committed to the war efforts in her novel, she does show how the divide affects her community.

Over the decade prior to 2012, when McBride started writing her novel, she served as a community college instructor and had military members in her class. "I had young men and women who were going off to Iraq and going off to Afghanistan, and coming back and taking classes sometimes when they were done, but more often, when they were in between deployments," McBride says. And because she taught composition to these students, she was able to see even more of their life experiences through their writing. In reading their work, she was struck by "the disconnect between the experiences they were having and the experiences of the other people in the class who were worrying about entirely different kinds of things." In her class were representatives of a nation at war, but only a few of them were directly involved.

Over time, McBride noticed a change in her students and in their writing. "There was this transition from enormous support for the war and pride in what they were doing and a willingness and an eagerness to talk about it," to something noticeably different when the national mood toward the war changed. "I remember it in my classes really well when there started to be this enormous tension" from her military students, many of whom were deploying for their third or fourth time. "There was a sense, again, that a very small amount of people were being asked to shoulder this burden that wasn't being shouldered, even remotely, by others," McBride says.

She captures this same transition in Nate, a central character in her novel. Nate, who had joined the military from Las Vegas, was initially proud of his service, even if he had seen some difficult things. But McBride says, "in my mind, it started to turn for him when he began to lose faith in who was making the decisions about what he would do." She was not consciously trying to mirror the transition in the perspectives of her students over the years, but looking back, she can see the connection.

Even though her novel matches her experience and evolving understanding of the effects of the war at home, McBride says that it was not the aim of her book to share a better understanding of the war with the civilian public. "I wanted to tell a story, and I wanted the reader to want to read it, but I did care about getting the story right," McBride says. Instead, she was writing a story she had learned from teaching her "students who were soldiers and the students sitting in the seats next to them" and how that told the story of her city.

Unlike McBride, Phil Klay was in fact trying to take part in a broader cultural conversation about the wars when he was writing *Redeployment.* When he came back from his tour in Iraq, he noticed "this sort of sense of something missing" from the dialogue about the war, and he tried "to write towards what that was." Writing allowed him "to solicit from other people a sense of what that might be," too, he says. "I don't think the answer is the book itself," he says, adding, "it's a more seriously emotionally engaged series of conversations about what the legacy of these wars is and continues to be." To him, "fiction is a place where people can meet, where they take the time to very seriously examine and think about the experiences of other people and also crucially about the sorts of moral decisions those characters are making."

Gallagher sees fiction as a way to better understand war as well. "It's often described as ironic that something as brutal as warfare produces great works of literature, but I think war is part of the human condition," he says. "Ignoring our impulse to destroy each other isn't going to make it go away, so it makes sense that people, whether they directly experienced war or not, seek to write about it and seek to understand it."

Beyond simply a way to introduce a broader conversation to better understand wars, these authors also see fiction as uniquely suited to go deeper than the normal conversations and understandings of the wars in Iraq and Afghanistan. Fiction does not only allow readers and writers to go beyond the basic narratives of the current politics, and it not only allows more people access to the important conversations that need to be had. Fiction also allows them to access and explore truths in different ways.

For Gallagher, fiction allows both readers and writers to explore truths from new perspectives and to gain personally from the process. It allows people to "empathize with people across the globe who may be similar to you," he says, "or maybe people two hours away who have a story that is very different from yours." Also, fiction "can provide a perspective that memoirs written so close to the moment generally can't." Memoirs "just don't have the emotional distance from the subject," and "they also don't have a deeper emotional texture that fiction does." To him, good literary fiction "demands, if not meaning, texture and resonance," it "needs to be the story not of the thing, but of the people, of the time, of that time's meaning and inheritance."

For Klay, fiction provides an avenue for understanding truths about war while providing them in ways that others could appreciate. Fiction allowed access to the intricacies of life-altering events, not just for himself as a writer, but for his audience as well. And it did so in a way that was lacking in much of the nonfiction about the wars.

Fiction allowed him to explore the cultural divide between those who served in the wars and those who wish to understand it beyond political narratives. "I think that one of the things about war, in America, especially today, is that the American relationship to war is one of incredible distance from it," he says, "yet narratives of war have such a primal power in this culture." They "mainline directly into a whole series of emotional reactions and understandings of American patriotism, masculinity, and all of these other things," he says, but because of the way the war and veterans are embraced for political reasons, it is often difficult to have a real conversation about them. "As soon as you open your mouth there are all of these suppositions that are present before you even begin talking."

The difficulty of being deployed and coming home to a nation that is embracing a different narrative of wartime service than those who experienced it do, is one of the reasons Klay wrote his book. He was interested in writing stories that were more than just works of literature, but were about things that were "deeply central to people's sense of self and the way that they are able to exist within a certain community." By writing his stories, and especially through fiction, he was not only able to examine the meaning of his own experiences in the life that followed, but also to do so in a way that allowed others to understand even if they did not share in his experiences or perspective.

Fiction "allows you to explore emotional truths," Klay says. "There is a curiosity people have about what it is like to go to war," he says, but the explanation of a single person's experience is "only going to be as good as the sensitivity and insight that they bring to analyzing it." Memoirs vary in depth, in motivation, and in many other ways, but fiction allowed Klay to pursue answers to the questions that were most interesting to him. And, because he wrote a short story collection, he was able to pursue them many different times, circling "around the themes from different angles and keep[ing on] attacking."

By quoting others, both McBride and Carpenter explain why they believe that fiction presents perspectives and truths that perhaps nonfiction cannot. McBride says, "from the time I was really young, my mother used to say, 'a novel was just as good of a place to get the truth as anywhere else.'" She believes that, and it has influenced her approach to her work. "I'm sure that I've read lots and lots about the war because I was interested in it," she says, "but I wasn't reading it thinking about a novel." In fact, she avoided reading nonfiction about the wars because she did not want it to influence her story. As an academic, she feels as though she has to facilitate a broad discussion with her students and accommodate many different perspectives. But when she writes fiction, she is attempting to use the story as she imagines it "to express what feels true."

Similarly, Carpenter relates a quote she heard that reflects the way fiction approaches truth. While on a panel discussing writing, another writer said, "journalism is in the business of uncovering the lies, and fiction is in the business of uncovering the truth." She also found that writing a work of fiction about the war gave her access that journalists would not have, especially when concerns of operational security kept journalists away from military bases after the raid that resulted in the death of Osama bin Laden. "I'll never forget the day I went, pregnant, to try and get my Navy base visit request fulfilled," she says. "Maybe because I was a woman and pregnant I was slightly less intimidating than a *New York Times* reporter." And the access went beyond the limits of the military bases as well. "I sat down with so many people over the course of my research," she says, and they allowed her to ask them personal questions about their lives, about their upbringing, and about their relationships. After she had those conversations with people, they were willing to discuss all kinds of details that, perhaps, they otherwise would not have.

The access that these authors and countless others like them were provided, either through their personal experiences or through interviewing others, provided them with an opportunity to give access to these wars to others. It allowed them to write works that expanded the conversation beyond the basic cultural and political narratives that are often attached to the wars in Iraq and Afghanistan, and to challenge many of those narratives directly at the same time. And it allowed them to capture these truths through fiction, which authors like Walt Whitman and Paul Fussell thought impossible, and in more authentic ways than had been provided through much of the wartime nonfiction that preceded them.

Conclusion

THE WARS in Iraq and Afghanistan are extraordinarily complex events. No matter your role in the war, either as a voice of opposition, a policy maker, a service member on the ground, or even someone who was not engaged, for you to gain a real understanding of all of the layers, factors, individuals, actions, and decisions is a very difficult, if not impossible, task. This is complicated by how information on the wars was presented and received, and the narratives that long mediated the facts.

The truth is that almost any comprehensive view on the wars, their causes, or the outcomes is destined to be flawed in some way. At some point, an observer might stop seeking new information, or draw an incorrect conclusion, and either way, future conclusions may be built on an inherent flaw. And these represent the best-case scenario, one where an individual truly wants to know the truth and is willing to change his/her mind based on new facts or a reconsidered position.

This book explores a case other than the best case, for the most part. It considers narratives that have become broadly accepted conventional wisdom, at least to certain constituencies. It considers narratives that have become accepted as facts by a large enough percentage of the population that future policy decisions can be made based on flawed perceptions, because those flawed perceptions are dominant.

Consider the three narratives covering the president's decision to go to war and stay there—that he unilaterally chose to invade Iraq, that he embraced a cowboy persona, and that he had no Phase IV plan. These are all parts of an anti-Bush master narrative that portrays President Bush as being arrogant, incurious, willing to take unnecessary risks on behalf of his nation, and possibly even motivated by protecting business interests over the nation's interest.

For those who wanted to believe in all or part of this master narrative, these narratives supported their preferred worldview.

For those who reject this master narrative, the president's invasion of Iraq, his tough rhetoric portraying *good* fighting the forces of *evil*, and his decision to surge troops into Iraq under the leadership of David Petraeus serve as comfortable and convenient counternarratives that combine to make a master narrative of their own. In this master narrative, Bush was fighting for the American way of life, he was willing to accept criticism because he was committed to a cause that was bigger than himself, and he was willing to make hard decisions that were not based on poll numbers or other standards of flexibility. To those who wanted or even needed to believe in something like American exceptionalism, or any number of other beliefs for that matter, Bush's decisions, actions, and rhetoric seemed heroic and consistent with their worldviews.

But there are other ways to understand the facts of these wars that go beyond these basic and simplistic political narratives. As this book attempts to show, Bush's decision to invade Iraq was enabled by many, at least some who may have had their own political interests in mind and who would later criticize his decision. And worse, those who enabled him to make the decisions that led to and kept his nation at war almost never took anything approaching responsibility for their involvement. Instead, they promoted narratives against the administration in efforts to stir political dissent, just as Bush and his team deployed narratives to build support for their war efforts.

This is the true power of narratives, particularly when they are used in periods of uncertainty of the magnitude presented by the wars in Iraq and Afghanistan. It is not merely that a narrative can tell a story that offers a counterargument or a new perspective that challenges the status quo. It is not merely that some may use language, rhetoric, and narratives to build energy and emotion toward their own political objectives. It is that narratives have the ability to change the focus from real policy discussions and hard decisions where the stakes are high, and to reframe the debate in terms that are momentarily convenient and potentially meaningless in the long term.

The latter narratives explored in this book offer a different perspective on the same topics. Instead of examining how policy decisions were framed at the highest levels of government, they look at how these policy decisions were framed by those who were on the ground, and even by those who were not personally involved in these wars but who wanted to understand a deeper truth about them. Here we see the real implications of decisions like the decision to invade Iraq or to surge more troops, and we find a different story altogether.

In "Heroes and Victims," this book attempts precisely this juxtaposition. It begins by exploring the major narratives characterizing service members and the impact the wars had on them with the views of the service members expressed in their own words. Not surprisingly, there is a significant difference between the conclusions drawn in each perspective. Those who embrace the narratives that align with the dominant political worldviews of the day see heroes and victims, while those who served in these wars saw something more complex: personal growth from the difficulty of their experiences, even if they had not yet fully processed it all.

In exploring the use of the phrase "chickenshit," this book attempts to show that a lot of the discontentment heard from those on the ground is as old as the military and even as old as combat. It also attempts to show how the concept of "chickenshit" serves as a master narrative in its own right for those who were deployed, allowing them to assign meaning to the decisions and indecision of the war, and to also write these things off as the negative outcomes of the politics that started it all and that kept it going.

The policy decisions made at the highest level of government had enormous impact on those on the ground. These decisions could prevent you from accomplishing your mission, they could prevent potential partners from wanting to take risks, and they could even get people killed. And worse, these decisions were nearly always shrouded in political narratives that stood in stark contrast to the reality on the ground. The deployed service member probably only had a passing interest in whether or not President Bush chose to invade Iraq unilaterally if she had to prepare for a mission or keep her fellow service members focused, ready, and alive. This narrative allowed for a simple and cynical understanding of what it all meant, and allowed them to focus on more important things.

The contrast between the two perspectives represented by the highest level of government and the deployed troops is enormous. One is operating in a world of political theater, while the other is operating in a world of sobering reality. Politics will be a part of every war, and simply hoping that it will not be is not a useful approach. However, expecting those who have been entrusted with the sacred decisions to send Americans to their potential death or injury to take seriously the impact of their power is not too much to ask.

Having an honest discussion about these wars, or any war, actually, is also not too much to ask. This is why it is interesting to see this undertaken almost exclusively by fiction writers. These authors certainly had worldviews and perspectives that shaped their understandings of the war before they began writing their novels and short stories. They certainly had beliefs, experiences, and opinions that could have led them toward easy political conclusions based on narratives that aligned nicely with their views. But they wanted something deeper. They wanted to challenge their own views, they wanted to tell a story that was real, and they wanted to have empathy for others, even those they disagreed with. They wanted to find an honest, authentic, and nuanced truth, and even better, they wanted to present that truth in a way that started a conversation with others, not shut down conversations.

An honest observer of these wars should want to mimic the approach of these fiction writers. This does not mean that everyone should have to write a work of literary fiction about these wars in order to better understand them. Rather, they should want to find the truth that goes beyond the received wisdom of narratives. They should be willing to be uncomfortable and even wrong. And they should be willing and even eager to engage with anyone, but especially those who they might disagree with.

This book attempts to show a way that a person can do exactly that, or at least to begin such an effort. It begins with the basic assumptions that narratives are a real part of our political discourse, and that they are enormously powerful. It assumes that all perspectives present in a debate will use narratives to pursue their political ends, and that doing so amounts to manipulation in order to drive an election or a policy outcome.

If some people want to know the truth, or even an approximation of the truth, they can begin with this basic assumption and understanding. They can next work to identify the prevailing narratives around a given subject. This book shows that narratives, at least those involved with matters of the magnitude explored here, have a number of characteristics. They often present a binary picture—they hate us for our freedom, or, every general prior to David Petraeus was a failure, and everything after him was a success, for example. They offer certainty in order to explain great uncertainty. They arrange the facts in a way that align with a specific worldview, often their own worldview, in order to build a following for their message. And, they are driven toward political outcomes that benefit the purveyors of the narrative.

Once the narratives are identified, a person can then begin informing himself or herself about the various narratives that exist to explain a certain decision or event. This means reading, listening, and exposure to all viewpoints, even opposing viewpoints, in order to see how the narratives play off of one another. It means noticing which facts are present and common among opposing narratives, and it means drawing a conclusion that takes all of these facts and factors into account.

The unsatisfying part of this approach is that it takes more work than there is often time, and in the end, there is no guarantee of certainty. But this is where the approach of the fiction writers covered in this book becomes so important. They sought to find a truth about these wars and to tell that truth in ways that were inviting to others. They told the truths they shared in order to engage with others from a perspective of empathy and respect for their views, and not in an adversarial way. This approach is inherently challenging to one's own deeply held views, which is valuable in its own right, but it also removes the binary aspect of narratives in which you are forced to either agree or disagree.

This is surely not the only approach to weeding through today's political discourse on a subject of this size and complexity. Others may disagree by offering other methods, while still others will simply just disagree. They might argue that this approach is naive in that it does not reflect the true nature of how politics works in contemporary America. Or they will claim that this approach begins from a defeatist position in that it accepts large-scale political manipulation and the prohibitive complexity of finding the absolute truth of these wars as facts. But a critical, humble approach is far superior to the arrogance of today's political discourse, a discourse that has resulted in devastating outcomes.

This approach assumes that the factors, personalities, and arguments that allowed for these wars and their outcomes truly are important. It takes seriously the need for real policy and political debate when the stakes are so high. And, it offers a way to understand how narratives allowed for many of the decisions and actions of these wars to take place, while providing a way to avoid the same missteps in the future, missteps that destroyed lives.

Bibliography

Bernardi, Daniel Leonard, Pauline Hope Cheong, Chris Lundry, and Scott W. Ruston. *Narrative Landmines: Rumors, Islamist Extremism, and the Struggle for Strategic Influence.* New Brunswick, NJ: Rutgers University Press, 2012.

Bin Laden, Usama (Osama). "Declaration of War against the Americans Occupying the Land of the Two Holy Places." In *Princeton Readings in Islamist Thought,* edited and introduced by Roxanne L. Euben and Muhammad Qasim Azman, 425–59. Princeton, NJ: Princeton University Press, 2009.

Bolger, Daniel P. *Why We Lost: A General's Inside Account of the Iraq and Afghanistan Wars.* Boston: Mariner Books, 2015.

Brand, Laurie. "The Islamic State and the Politics of Official Narratives," Washington Post, September 8, 2014. https://www.washingtonpost.com/news/monkey-cage/wp/2014/09/08/the-islamic-state-and-the-politics-of-official-narratives/?utm_term=.eb215f90422a.

———. *Official Stories: Politics and National Narratives in Egypt and Algeria.* Stanford, CA: Stanford University Press, 2014.

Bremer, L. Paul. *My Year in Iraq: The Struggle to Build a Future of Hope.* New York: Simon and Schuster, 2006.

Broadwell, Paula, and Vernon Loeb. *All In: The Education of General David Petraeus.* New York: Penguin Press, 2012.

Bush, George W. *Decision Points.* New York: Broadway Paperbacks, 2010.

Buzzell, Colby. *My War: Killing Time in Iraq.* New York: Berkley Caliber, 2005.

Carpenter, Lea. *Eleven Days.* New York: Alfred A. Knopf, 2013.

Casey, George. *Strategic Reflections: Operation Iraqi Freedom, July 2004–February 2007.* Washington, DC: National Defense University Press, 2012.

Cheney, Dick, and Liz Cheney. *In My Time: A Personal and Political Memoir.* New York: Threshold Editions, 2011.

Devji, Faisal. *Landscapes of the Jihad: Militancy, Morality, Modernity.* Ithaca, NY: Cornell University Press, 2005.

Earle, Robert. *Nights in the Pink Motel: An American Strategist's Pursuit of Peace in Iraq.* Annapolis, MD: Naval Institute Press, 2008.

Fanning, Rory. "Army Veteran: Why Do We Keep Thanking the Troops for Their 'Valor'?" AlterNet.org, October 26, 2014. https://www.alternet.org/visions/army-veteran-why-do-we-keep-thanking-troops-their-valor.

Feith, Douglas J. *War and Decision: Inside the Pentagon at the Dawn of the War on Terrorism*. New York: Harper, 2008.

Fink, Naureen Chowdhury, and Benjamin Sugg. "A Tale of Two Jihads: Comparing the Al-Qaeda and ISIS Narratives." International Peace Institute *Global Observatory*, February 9, 2015. https://theglobalobservatory.org/2015/02/jihad-al-qaeda-isis-counternarrative/.

Fisher, Louis. *Presidential War Power*, 2nd ed., rev. Lawrence: University Press of Kansas, 2004.

Frankfurt, Harry G. *On Bullshit*. Princeton, NJ: Princeton University Press, 2005.

Fussell, Paul. *Wartime: Understanding and Behavior in the Second World War*. New York: Oxford University Press, 1989.

Gallagher, Matt. *Kaboom: Embracing the Suck in a Savage Little War*. Cambridge, MA: Da Capo Press, 2010.

———. *Youngblood*. New York: Atria Books 2016.

Gates, Robert M. *Duty*. New York: Alfred A. Knopf, 2014.

Gentile, Gian. *Wrong Turn: America's Deadly Embrace of Counterinsurgency*. New York: New Press, 2013.

Gordon, Michael R., and General Bernard E. Trainor. *Cobra II: The Inside Story of the Invasion and Occupation of Iraq*. New York: Pantheon Books, 2006.

Haass, Richard N. *War of Necessity, War of Choice: A Memoir of Two Iraq Wars*. New York: Simon and Schuster, 2009.

Halverson, Jeffry R., H. L. Goodall Jr., and Steven R. Corman. *Master Narratives of Islamist Extremism*. New York: Palgrave Macmillan, 2011.

Hartley, Jason Christopher. *Just Another Soldier: A Year on the Ground in Iraq*. New York: Harper, 2005.

Hoffman, Cara. "Stop Calling Soldiers 'Heroes': It Stops Us from Seeing Them as Human—and Dismisses Their Experiences." Salon.com, July 20, 2014. https://www.salon.com/2014/07/20/stop_calling_soldiers_heroes_it_stops_us_from_seeing_them_as_human_and_dismisses_their_experience/.

Holyfield, Lori. *Veterans' Journey Home: Life after Afghanistan and Iraq*. Boulder, CO: Paradigm Publishers, 2011.

Isikoff, Michael, and David Corn. *Hubris: The Inside Story of Spin, Scandal, and the Selling of the Iraq War*. New York: Broadway Books, 2007.

Kagan, Kimberly. *The Surge: A Military History*. New York: Encounter Books, 2009.

Kaplan, Fred. *The Insurgents: David Petraeus and the Plot to Change the American Way of War*. New York: Simon and Schuster, 2013.

Klay, Phil. *Redeployment*. New York: Penguin Books, 2014.

Lacey, Jim, ed. *A Terrorist's Call to Global Jihad: Deciphering Abu Musab Al-Suri's Islamic Jihad Manifesto*. Annapolis, MD: Naval Institute Press, 2008.

Mansoor, Peter R. *Surge: My Journey with General David Petraeus and the Remaking of the Iraq War*. New Haven, CT: Yale University Press, 2013.

Masciotra, David. "You Don't Protect My Freedom: Our Childish Insistence on Calling Soldiers Heroes Deadens Real Democracy." Salon.com, November 9, 2014. https://www.salon.com/2014/11/09/you_dont_protect_my_freedom_our_childish_insistence_on_calling_soldiers_heroes_deadens_real_democracy/.

McBride, Laura. *We Are Called to Rise.* New York: Simon and Schuster, 2014.

Minear, Larry. *Through Veterans' Eyes: The Iraq and Afghanistan Experience.* Washington, DC: Potomac Books, 2010.

Owens, Mackubin Thomas. "Life after Wartime: Combating the Veteran-as-Victim Narrative." *Weekly Standard,* June 2, 2014, 21–24.

Porch, Douglas. *Counterinsurgency: Exposing the Myths of the New Way of War.* Cambridge: Cambridge University Press, 2013.

Rice, Condoleezza. *No Higher Honor: A Memoir of My Years in Washington.* New York: Crown, 2011.

Ricks, Thomas E. *Fiasco: The American Military Adventure in Iraq.* New York: Penguin Books, 2007.

———. *The Gamble: General David Petraeus and the American Military Adventure in Iraq, 2006–2008.* New York: Penguin Press, 2009.

Robinson, Linda. *Tell Me How This Ends: General David Petraeus and the Search for a Way Out of Iraq.* New York: PublicAffairs, 2008.

Rudd, Gordon W. *Reconstructing Iraq: Regime Change, Jay Garner, and the ORHA Story.* Lawrence: University Press of Kansas, 2011.

Rumsfeld, Donald. *Known and Unknown: A Memoir.* New York: Sentinel, 2011.

Salazar, Philippe-Joseph. *Words Are Weapons: Inside ISIS's Rhetoric of Terror.* New Haven, CT: Yale University Press, 2017.

Schultz, Howard, and Rajiv Chandrasekaran. *For Love of Country: What Our Veterans Can Teach Us about Citizenship, Heroism, and Sacrifice.* New York: Alfred A. Knopf, 2014.

Special Inspector General for Iraq Reconstruction. *Hard Lessons: The Iraq Reconstruction Experience.* Washington, DC: US Government Printing Office, 2009.

Stephenson, James. *Losing the Golden Hour: An Insider's View of Iraq's Reconstruction.* Washington, DC: Potomac Books, 2007.

Summers, Benjamin. "Hero Worship of the Military Is Getting In the Way of Good Policy." *Washington Post,* June 20, 2014. https://www.washingtonpost.com/opinions/hero-worship-of-the-military-presents-an-obstacle-to-good-policy/2014/06/20/053d932a-f0ed-11e3-bf76-447a5df6411f_story.html?noredirect=on&utm_term=.66fc62023125.

Tupper, Benjamin. *Greetings from Afghanistan, Send More Ammo: Dispatches from Taliban Country.* New York: NAL Caliber, 2010.

Van Buren, Peter. *We Meant Well: How I Helped Lose the Battle for the Hearts and Minds of the Iraqi People.* New York: Metropolitan Books, 2011.

Wolfe, Tom. "Stalking the Billion-Footed Beast." *Harper's* (November 1989). https://harpers.org/archive/1989/11/stalking-the-billion-footed-beast/.

Wood, Graeme. "What ISIS Really Wants." *Atlantic* (March 2015). https://www.theatlantic.com/magazine/archive/2015/03/what-isis-really-wants/384980/.

Wood, Trish. *What Was Asked of Us: An Oral History of the Iraq War by the Soldiers Who Fought It.* New York: Little, Brown, 2006.

Woodward, Bob. *Bush at War.* New York: Simon and Schuster, 2002.

———. *Plan of Attack.* New York: Simon and Schuster, 2004.

———. *State of Denial.* New York: Simon and Schuster, 2006.

———. *The War Within: A Secret White House History, 2006–2008.* New York: Simon and Schuster, 2008.

Index